Integration in Energy and Transport

Contemporary Central Asia: Societies, Politics, and Cultures

Series Editor: Marlene Laruelle, George Washington University

At the crossroads of Russia, China, and the Islamic world, Central Asia remains one of the world's least-understood regions, despite being a significant theater for muscle-flexing by the great powers and regional players. This series, in conjunction with George Washington University's Central Asia Program, offers insight into Central Asia by providing readers unique access to state-of-the-art knowledge on the region. Going beyond the media clichés, the series inscribes the study of Central Asia into the social sciences and hopes to fill the dearth of works on the region for both scholarly knowledge and undergraduate and graduate student education.

Titles in the Series

Kyrgyzstan beyond "Democracy Island" and "Failing State": Social and Political Changes in a Post-Soviet Society, edited by Marlene Laruelle and Johan Engvall
Afghanistan and its Neighbors after the NATO Withdrawal, edited by Amin Saikal and Kirill Nourzhanov
Integration in Energy and Transport: Azerbaijan, Georgia, and Turkey, by Alexandros Petersen

Integration in Energy and Transport

Azerbaijan, Georgia, and Turkey

Alexandros Petersen
Foreword by Roy Allison

LEXINGTON BOOKS
Lanham • Boulder • New York • London

Published by Lexington Books
An imprint of The Rowman & Littlefield Publishing Group, Inc.
4501 Forbes Boulevard, Suite 200, Lanham, Maryland 20706
www.rowman.com

Unit A, Whitacre Mews, 26-34 Stannary Street, London SE11 4AB

British Library Cataloguing in Publication Information Available

Library of Congress Cataloging-in-Publication Data Available

ISBN: 978-1-4985-2553-4 (cloth : alk. paper)
ISBN: 978-1-4985-2554-1 (electronic)

∞™ The paper used in this publication meets the minimum requirements of American National Standard for Information Sciences Permanence of Paper for Printed Library Materials, ANSI/NISO Z39.48-1992.

Printed in the United States of America

Table of Contents

Foreword

This book offers an original exploration of integration, in the energy and transport sectors, among Azerbaijan, Georgia, and Turkey and the capacity of this to change the relations between these countries. Alexandros Petersen studies in particular the way transnational extra-regional actors promote such integration. These actors include European Union and World Bank projects designed to improve transparency and governance in the energy sector and to promote transport through the South Caucasus and Turkey. He finds intriguing complementarities in aspects of relations between these three countries, which are often described as forming a common and strategically significant energy and transport "corridor" westward. International institutions, he argues, appear to be catalyzing processes of integration. But in the period studied, mid-1990s to 2008, such integration significantly did not "spill over" into economic, security, or foreign policy more generally. Petersen sets his analysis in a theoretical framework, drawing on theories of integration, but he also grounds it in the detailed, empirical knowledge that is the measure of true expertise.

Petersen developed a reputation in the late 2000s as a young, even precocious, analyst and scholar, displaying wide-ranging interests and building up a growing portfolio of policy-related publications and affiliations with research projects. He was increasingly prominent as an articulate commentator for a wide variety of research bodies and think tanks as well as the media in Western Europe, the United States, the South Caucasus, and Central Asia. The positions he held speak to his interests. To list a few: Senior Fellow with the Eurasia Center and Fellow for Transatlantic Energy Security at the Atlantic Council; Southeast Europe Policy Scholar and, later, Advisor to the European Energy Security Initiative at the Woodrow Wilson Center, Washington, DC; Adjunct Fellow with the Russia and Eurasia Program at the Center for Strategic and International Studies, Washington, DC; Program Director of the Caspian Europe Center in Brussels; and Associate Professor at the American University of Central Asia. His work was focused on "Eurasia," defined as the geographic space centered on the vast region from the Black Sea to Chinese Central Asia. He developed a passionate commitment to study and explain to various audiences the tectonic shifts underway in Eurasian geoeconomics and geostrategy and to explore their implications for Western states.

However, Petersen's determination to do so through direct knowledge, through the friendships he cultivated and fieldwork, undaunted by the rigors of travel, brought him to Kabul in early 2014. Here by awful chance his career and short intrepid life was cut down in a shocking attack by the Taliban, just days after he began a new academic attachment as an assistant professor in political science at the American University of Afghanistan. He was twenty-nine years old. When we exchanged messages earlier that week I congratulated him on his new post, while he looked forward to hosting me in his new academic setting: "if you find yourself traveling through Kabul." It was not to be, and a dynamic and original voice was stilled. However, Petersen's publications will continue to influence those drawn to understanding the big picture of the region where he died, the interplay of energy, security, and politics, as well as the meshing and conflicting interests of major powers and small vulnerable states.

This volume was originally written by Petersen as a doctoral thesis for the London School of Economics, at a time when the South Caucasus had established itself as a corridor for energy, trade, and associated infrastructure from Azerbaijan, via Georgia to Turkey and on to Europe, symbolized by the Baku-Tbilisi-Ceyhan oil pipeline. The new infrastructure being developed was creating an east-west "Eurasian bridge." Competing schemes promoted by Russia for north-south energy transportation via the Caspian Sea remained unpromising, even if Russia had locked in its own energy routes from the north of the Caspian Sea, via the North Caucasus and the Black Sea.

Petersen saw energy and transit as facilitators of an organic process of integration, but also was keenly aware from the outset that the states of the Western-oriented corridor were diverse in their political systems and foreign and security policies; they lacked common attitudes to their neighbors, cultural commonalities, or clearly complementary goals in security policy. There had been some attempts at trilateral cooperation in limited fields of soft security, such as pipeline security, but broader institutionalized regional cooperation did not seem in prospect. His research confirms that political, security, and sociocultural integration between Azerbaijan, Georgia, and Turkey have lagged far behind energy and transport integration in the region; indeed, the extent to which this constitutes a separate "region" in any broader sense is questionable.

For Petersen the key to understanding the particular interactions in energy and trade among the three states is first that energy and transport are both highly regulated sectors which lend themselves to standardization across borders. Secondly and crucially these sectors have engaged multinational corporations and supranational institutions. What emerges as the central explanatory variable, which accounts for the integration observed, is the contribution of *transnational extra-regional actors*, especially the prompting of the European Union and international financial institutions. These bodies have provided technical assistance and financial support to integrative projects. They have pursued broad agendas beyond simple profit, such as encouraging governments to ensure transparency of finances and to respect contracts with private partners.

Petersen studies two main case studies. First, the EU-sponsored projects Transport Corridor Europe-Caucasus-Asia (TRACECA) and Interstate Oil and Gas Transport to Europe (INOGATE). Secondly, the World Bank's efforts at promoting harmonization and transparency in the energy sector. These detailed studies provide the empirical core of his argument. However, to confirm the claim that transnational extra-regional actors have acted as the key variable promoting integration, another counterfactual case study is offered, that of the natural gas pipeline network from Turkmenistan to China—a project that received no technical or financial assistance from transnational extra-regional actors. Petersen duly finds that there is little evidence of integration among the transit countries, Kazakhstan, Turkmenistan, and Uzbekistan, and that this grand pipeline system has not been accompanied by greater integration in the energy sector or other sectors in these states.

However, the particular and restricted integration in the energy sector among the South Caucasus corridor states also reflects the political systems in place. Azerbaijan, the producer and exporter of energy, has exerted especially strong influence on regional energy policy. Energy integration has taken the form of a top-down process in Azerbaijan's highly centralized political system and has had little impact on other realms of policy making in Baku. This state-driven process is quite far from the bottom-up regionalism, sometimes referred to as soft regionalism, found in other world regions, which has been driven by civil society, NGOs, and small businesses. Petersen shows how integration in transport infrastructure has been a much more decentralized and public affair, as it affects various other fields such as immigration, customs, tariffs, and trade policy.

Overall, Petersen is cautious with his conclusion that the integration so far between Azerbaijan, Georgia, and Turkey in the fields of energy and transport is "not necessarily a unidirectional or progressive process," but it has helped to forge a limited bond that unites the three countries to the exclusion of their neighbors. It is an emerging if limited community of states that have little previous history of collaboration, but have found mutual benefit in cooperative arrangements. This conclusion continues to be borne out even as the geopolitical headwinds of the Russian annexation of Crimea and military intervention in east Ukraine push and pull the states between the Caspian and Black seas. Indeed in some sense the technical co-operation over energy and transport between Azerbaijan, Georgia, and Turkey is even more important as the glue of regional interstate cooperation at a time when taking firm positions on security policy, or political partnerships with the EU, NATO, or the United States could draw these states into the wider confrontation between Russia and Western states. At the same time many developments have reinforced the significance of the east-west energy and transport corridor: these include the collapse of Russia's plans for an alternative "South Stream" pipeline route, which would have competed with oil and gas pipelines through the South Caucasus; the emergence of China's grand "Silk Road" transport and infrastructure scheme, which will construct railway and freight transport connections all the way from China's western provinces, through Kazakhstan, Turkmenistan, over the Caspian Sea and via Azerbaijan westward to Turkey; and the prospect of Iranian energy reentering the market, among other routes possibly westward across the Caspian Sea. However, geopolitical vulnerabilities remain: Azerbaijan continues to precariously balance its external relations between Russia and Western states; Georgia, the critical transit state, still absorbing the effects of the August 2008 war with Russia, fears another more fundamental Russian intervention; while Turkey's volatile domestic politics could result in unexpected reorientations in its external relations.

Petersen would have relished continuing his analyses of the shifting tectonic plates of contemporary Eurasia, despite the evidence of increasing polarization in regional politics and the entrenchment of authoritarian rule. He set out his stall in his previous book, *The World Island: Eurasian Geopolitics and the Fate of the West*, published in 2011. This volume explores grand strategy through the visions of twentieth-century Eurasian geopolitical thinkers and proposes a new synthesis of ideas to reestablish Western strategic objectives. It studies the potential for Western engagement with China, Afghanistan, Turkey, Russia, and other Eurasian states. To promote Western values Petersen presented a strategy for the development of trade and energy links, coupled with the promotion of good governance, integration, and Western orientations among the Eurasian nations.

It is far too early to conclude whether such a strategic vision can gain traction, as Western states struggle with Islamist threats in the Middle East and North Africa, and with concomitant refugee flows into the European Union, while Russia continues to flaunt the use of hard military power in Ukraine. However, Petersen was prescient in seeking to focus attention on the growing strength of China as a Eurasian land power and trading giant. Indeed at the time of his death he was underway with another book, on China and Central Asia, with his coauthor Rafaello Pantucci.

For Petersen the Azerbaijan-Georgia-Turkey energy and transport corridor was indispensable in advancing any wider Western strategy of engagement with Eurasia. He recognized the need for the carefully researched and documented study that follows. He filled the gap and it forms a central part of his intellectual and scholarly legacy. However, his contributions to the analysis and understanding of Eurasian states and regions would certainly have won still greater acclaim if his life had not been taken so pointlessly by those who understood nothing of his fascination with and commitment to the people and places he studied.

Roy Allison
Professor of Russian and Eurasian International Relations, University of Oxford
Director, Russian and Eurasian Studies Centre, St. Antony's College, Oxford
September 2015

Introduction

A limited process of integration has been occurring among the countries of Georgia, Azerbaijan, and Turkey. From the mid-1990s to 2008, integration among the three countries has occurred in the energy and transport sectors, but not in other sectors, such as security, politics, or trade. In the energy sector, this integration can be explained through neoliberal institutionalist theory. Integration in the transport sector has occurred due to a mixture of elements from the neoliberal institutionalist and security communities and neofunctionalist theories of integration. In both sectors, transnational extra-regional actors (TERAs) are the explanatory variable in the integration. These theories will be analyzed in greater depth below.

This process occurred during this book's time period of study (mid-1990s to 2008) despite sometimes flaring tensions between Russia and Georgia, which led to a brief but hot conflict in 2008, the ongoing Nagorno-Karabakh conflict between Azerbaijan and Armenia, and a history of noncooperation between Georgia, an Orthodox Christian country, and its Turkic Islamic neighbors.[1] This integrative trend possesses characteristics distinct from analogous processes in regions such as Western Europe and Southeast Asia, which have helped to transition constellations of small countries into dynamic regional trading blocs to their mutual benefit.[2] First, the integration among Azerbaijan, Georgia, and Turkey is largely restricted to the high-value and highly regulated spheres of energy and transport. It has not spilled over into political and economic integration writ large. In part, this may be a self-perpetuating phenomenon: as the energy and transport sectors are treated differently because they are harmonizing, they will continue to be treated differently in the future. This is not the entire story. Highly regulated sectors such as these lend themselves to standardization and therefore harmonization when it is in the interests of the countries to do so. Energy resources, such as oil and natural gas in Azerbaijan, are extremely valuable and governed outside the legal-rational state apparatus. Consequently, energy as a sector is treated as sui generis by the countries of the region.

In addition to this major distinction, the countries of Georgia, Azerbaijan, and Turkey possess another difference as well: the organic integration that is occurring among these three countries is *cultivated,* that is, it is orchestrated as a top-down process by international institutions operating in the region in pursuit of a broader agenda.[3] Identified here as transnational extra-regional actors (TERAs), these institutions have provided financial, material, and legal support for the harmonization of the transport and energy sectors not solely for purposes of resource extraction and profits, but also for purposes of transparency, proper accounting of state finances, good governance, and most importantly a predictable business environment that is attractive to foreign investors.

This process of integration, even limited to the energy and transport sectors, is multifaceted and not necessarily progressive or even unidirectional. Nor are the two sectors themselves proceeding in the same manner. The energy sector, particularly in Azerbaijan, is a shadowy realm governed from the highest levels of government, outside the reach of ordinary parliamentary legislation and outside the political constraints of the ordinary electorate. TERAs and other international institutions that attempt to influence energy policy in Azerbaijan must do so from the very top—it is here that production and extraction agreements with foreign oil companies are negotiated; here where state oil profits are divided and dispersed; and here where major foreign policy decisions concerning the export of Azerbaijani oil and gas are made.[4] By sharp contrast, the transport and communications sector (treated as one for the moment because the two components possess similar structural characteristics) affects ordinary lives, shapes broad social transformations, and engages highly public regulatory reform (such as harmonization of customs and border controls), public infrastructure financing (such as the sale of bonds), and tariff and trade policy. This is not to overstate the binary distinction—pipeline security, for instance, is an aspect of energy integration that affects local control and requires local cooperation—but nonetheless, two separate processes are underway. While integration in the energy sector is almost entirely cultivated by TERAs and international financial institutions, integration in the transport and communications sector is a more diffuse process.

Why has this sector-specific integration not led to "spillover" in other sectors, to borrow the word, analyzed further below, from the neofunctional theory that integration is like a row of dominoes that accelerates its own continued momentum? There is no easy answer to this question, but the dynamics of top-down energy reform, cultivated by TERAs and other donors, may simply have limited impact on other sectors. Azerbaijan covetously guards its energy resources and views them as integral to its national prestige. Perhaps by contrast, the transport sector may be seen as a highly technical sphere and one that does not affect national security interests.[5] In any event, the two sectors appear to be treated autonomously and separately from other forms of policy making, which blunts the broader consequences of their emerging integrative tendencies. These themes form the core of this book and will be drawn out in the examination that follows.

WHAT IS INTEGRATION?

Unlike the traditional realist assumption that a state always acts, and only acts, in its own rational self-interest, integrative theory conceives of national interests in mutable and highly contingent ways and attempts to explain why, especially after the Cold War, former rivals have sought to cooperate with one another even while doing so appears to be at odds with national security and hard-power interests. The earliest steps of European integration in the 1950s launched the field of integration theory because they pointed to "one of the most significant challenges to realism's dominance, since they offered a view of international politics that stressed precisely how anarchy might be transcended, and a notion of politics that focused on governance and consent rather than power relations between autonomous and atomised individual actors."[6]

At the same time, it would be a logical fallacy for theorists of integration in international relations to assume that the most successful case—the integration of Western Europe—is necessarily representative of integration as a theoretical process. When Chryssochoou, for instance, describes European integration as a process of joining together "distinctive political units through a politics of accommodation, institutionalized compromise, and an informal culture of consensus building at the highest political level," he is identifying characteristics that may be peculiar to the European case.[7] While cooperation and consent are required for the definition of integration as distinct from coercive processes, the type of governing politics is not strictly necessary.

The mechanism by which integration occurs has long been debated by international relations theorists. In general, the three main theories posit that integration occurs when popular expectations about national self-identity change (security communities theory); when regulatory and administrative harmonization at low levels leads to cooperation at higher levels (neofunctionalism); or when international institutions cultivate and perpetuate integration among their member states, with their own organizational agendas in mind (neoliberal institutionalism). Although each of these three theories identifies a different triggering mechanism, and a host of additional hybrid theories attempt to divide or combine permutations of them, the three theories have a common core: the national interests of states begin to converge as a result of their cooperation, and the convergence of national interests changes state behavior accordingly.

The first of these traditional theories of integration is security communities theory, pioneered by Karl Deutsch in describing the integration of Western Europe in the late 1950s. According to Deutsch and his colleagues, integration is a relationship among political units in which they are mutually interdependent and jointly possess system properties that they individually lack.[8] As states delve deeper into cooperative arrangements, the prospect of open hostilities between them declines because information is more easily and more widely shared and because membership in a common community increases the predictability of state behavior. As the populations of these nation-states engage in increasing interactions, economic transactions, and communications with one another without regard to borders, their self-identities begin to converge and a collectivist identity emerges. The value of this theory is that it draws attention to actors other than the nation-state by focusing on soft-power cooperation such as sports, academic exchange, technology transfer, and media broadcasting. These points of contact, taken in the aggregate, produce a kind of "we" feeling" or shared sense of community. Security communities theory faced a revival with the constructivist revolt against realism after the Cold War. Constructivist theory in international relations posits that anarchy among nation-states is learned, and can be unlearned as states increasingly grow to share a collective identity through discourse and social practices.[9] Rooted in the human sciences, such as sociology, constructivism does not take state interests for granted but recognizes that they can change depending on the context.[10]

The second of the traditional theories is neofunctionalism, which was developed by the scholars Ernst B. Haas and Leon Lindberg, following the creation of the European Coal and Steel Community (ECSC) in the late 1950s; they sought to explain how ECSC's regulatory harmonization efforts had "spilled over" into broader economic integration (through the European Economic Community) and energy integration (through Euratom).[11] According to neofunctionalism, as interest groups begin to turn toward supranational communities to realize their demands they shift loyalties to the supranational institution, and under the pressure of these groups loyal to the supranational organization, the removal of simple technical barriers such as currency and labor controls would accelerate and beget broader and deeper political union. Neofunctionalism ran into difficulty in the early 1960s, however, as European integration stalled for several years due to political difficulties in countries like France, undermining the central tenet of the theory that the integrative process, once it began, would gain momentum and become self-reinforcing. However, although the traditional concept of spillover is now recognized as being unique to the European context and not necessarily transferable to other regions of the world, neofunctionalism as a partial theory also gained some traction following the end of the Cold War.[12] The theory operates more as a lens than as a comprehensive framework: by focusing on special interest groups, technocratic change, and low-level cooperation by nonpolitical elites, the neofunctionalist theory of integration is able to account for a more diverse array of actors than state-centric theories are.

The final traditional theory of integration in international relations is neoliberal institutionalism, proposed in its current incarnation by Robert Keohane, who argued that nation-states will form collective institutions to achieve their own interests, but once built these institutions will alter state behavior by establishing and enforcing norms, sharing information, and reducing transaction costs.[13] Participation in collective institutions can help a state build its reputation among other member states and can increase the costs of defection or noncompliance. Unlike the other traditional theories, neoliberal institutionalism accepts the realist assumption that states are the primary actors in international relations. Realism may well agree with the proposition that states will join collective arrangements if it is in their national interests to do so; where neoliberal institutionalism departs from realism, however, is by arguing that participation in collective organizations will change state behavior and ultimately change those national interests.

The three theories will be helpful in analyzing the progress of integration in the energy and transport sectors in Azerbaijan, Georgia, and Turkey. In particular, a major focus of this integrative process is the important role played by international actors with broad-based agendas for reform, here called TERAs, which can reach the highest levels of government, reduce the costs of compliance and increase the costs of defection, and eventually alter popular expectations. TERAs have had particular success in integrating the transport sector (and, perhaps derivatively, the communications sector) and the energy sector, two highly regulated and financially valuable sectors that lend themselves to cross-border transactions in order to reach regional and global markets. The success of TERAs outside these sectors is very limited, and these sectors have been segmented from the rest of the economy (and, in the case of energy in Azerbaijan, insulated from the ordinary political process).

The failure of integration outside the energy and transport sectors among Azerbaijan, Georgia, and Turkey belies the traditional neofunctionalist concept of spillover in which integration in one regulatory sphere would create pressure to integrate related spheres. In the security sector, for instance, Turkey and Azerbaijan were unable and unwilling to come to the aid of Georgia when hostilities flared with Russia in August 2008. The security relations of each country have dramatically altered depending on the leader in charge; Azerbaijan and Georgia are still dependent in some ways on Russia, and will continue to be. Even isolation of Armenia, supported by Azerbaijan and Turkey after the war in Nagorno-Karabakh in 1994, is only lukewarmly embraced by Georgia. Similarly, in the political sphere, all three countries have ambiguous relations with Russia, with Georgia's relationship the most tense. Azerbaijan, largely dependent on oil production for its national wealth, is in some danger of "Dutch disease," which in the economic sense refers to the decline in productivity in other sectors once an economy becomes overly dependent on a valuable extractive resource, which makes the currency so strong that exports become noncompetitive.[14] In addition, Azerbaijan's political system is not fully rational, and the democratization process in Georgia following the election of Mikhail Saakashvili in 2004 is still young. It may be partly for these reasons that the security and political sectors have shown very little evidence of integration, but that is not the focus of this study.

Likewise, the cultural and economic spheres have also shown virtually no evidence of integration among the three countries to the exclusion of their neighbors. In the cultural sphere, the three countries can only count to two: two speak a Turkic language and favor largely secular Islam, while one is Orthodox Christian; two have a common history of Soviet Russian domination and are currently experiencing a revival of their national languages; and two have a generally pro-European sociopolitical orientation and wish to join the European Union. The three countries together have fairly little in common on a cultural level that they do not share with their neighbors. On an economic plane, integration is multidirectional and uneven. Both Azerbaijan and Georgia are not completely economically independent of Russia, and even Turkey counts Russia as a top trade partner after the European Union. The best-integrated country of the three is probably Georgia, given its location as a transit center, but its share of trade is only a fraction of Azerbaijan's and Turkey's. Travel among the three countries can be difficult, and barriers exist to labor migration among the three. In addition, the informal sector, particularly illicit trade in drugs and other regulated contraband, as well as some reports of terrorist activities, undermines economic integration to some degree. The three countries do not possess entirely complementary economic agendas and face different market pressures.

Integration among Azerbaijan, Georgia, and Turkey is occurring because all three countries stand to benefit from cooperation, as each possesses a resource that the other two lack on their own. In this, they have a natural complementarity. For Azerbaijan that resource is oil, a high-demand and high-value good that requires transport to world markets. Georgia, meanwhile, is able to offer a transit corridor (including Black Sea ports) that is strongly linked to regional markets and provides the only inlet into Armenia. Turkey, for its part, possesses links to global markets, including continental Europe, and strong relationships with international donor countries and financial institutions. The three countries together are more than the sum of their individual parts, which is fueling integration in those very sectors in which their complementarity is highest: namely, energy and transport. This may also explain the lack of progress in other sectors: they simply lack the structural incentives for integration that exist in the energy and transport sectors.

A DEFINITION OF INTEGRATION

According to international relations theorist Joseph Nye, integration involves (a) at least rudimentary institutional structures among the countries involved; (b) at least some interdependence in policy formation and implementation; and (c) at least some collective sense of mutual identity and obligation.[15] Although Nye is, with Keohane, traditionally considered the cofounder of neoliberal institutionalism, his definition includes observable characteristics that encompass not only a neoliberal institutionalist characteristic (existence of institutional structures), but also a neofunctional quality (policy harmonization) and a constructivist or identity-based quality. The existence of a supranational framework possessing a separate and distinct personality from member states can strengthen integrative-process forces and improve compliance.[16] International institutions, however, do not universally improve identity formation or distributional equity among member states, and so judgments must be made on a case-by-case basis. According to Nye, political integration has observable characteristics. On an institutional level, the expansion of capacity in the form of budget and staffing and the expansion of legal jurisdiction can signify integration. On a policy level, the scope of the policy- and the decision-making locus are likewise measurable. Attitudinal changes can be tracked through elite and public polling and transactions and exchanges across borders. Finally, the absence of hostile incidents, as explained by Deutsch, could signify the emergence of a security community or a non-war zone.[17]

Integration among Azerbaijan, Georgia, and Turkey is an appropriate case study for this definition of integration. The international institutions involved, TERAs, possess a characteristic of extra-regionality in the sense that they are not based in the region or even necessarily include the three countries under study here as formal members. Nonetheless, TERAs such as the European Union and the World Bank are able to create a framework in which the three countries can predict each other's behavior, trust in one another's compliance with the institutional norms, and lower the transactional costs of repeatedly engaging one another. At least in the energy and transport sectors, Azerbaijan, Georgia, and Turkey have repeatedly negotiated policy formation that led to such high-cost projects as the Baku-Tbilisi-Ceyhan (BTC) pipeline, updated rail networks, and harmonization of customs duties. Finally, the evidence may show a change in attitudes and expectations, particularly in the transport sector which (compared to energy) is a sector engaged with a broad slice of the population and not limited to political elites. Because of the natural compatibility of Azerbaijan, Georgia, and Turkey, with each possessing different comparative advantages—Azerbaijan with resources, Georgia with its transport corridor, and Turkey with access to world markets—the involvement of TERAs in the transport and energy sectors are helping to solidify and accelerate integrative process forces as predicted by Nye.

Integration theory attempts to fill a traditional gap in the realist paradigm of international relations, which holds that states act in their own self-interest, and as those interests change state behavior will follow.[18]

Keohane and Nye, however, writing in the early 1970s, saw a more complex pattern of actors and issues evolving, at least among industrialized nations, that tended to reduce the importance of military and security matters.[19] The three theories under study here each identify a causal mechanism that is traditionally not included in the realist toolbox. For Deutsch and security communities theory, popular communications and transactions with a neighboring state can shape the attitudes and ultimately the behaviors of states toward one another. For Haas's neofunctionalism, supranational elites such as business leaders and unions would work toward regulatory harmonization that would place pressure on other sectors to integrate and ultimately create self-sustaining momentum toward integration. Finally, for Keohane, building on Nye's foundation, international institutions proved to be a decisive variable in altering state interests and eventually behavior.

Integration, as defined by this study, is not universally a "good" phenomenon, despite the efforts by Deutsch and other scholars to emphasize the relative peacefulness of states that develop commonalities with their neighbors. In actuality, integration produces changes and stresses that can enhance or diminish values, have unwanted distributional consequences, or create conflict between the center and periphery.[20] It is an agnostic and an amoral process. Similarly, integration is likely a reversible process, and one may expect that as the interests of nation-states become unaligned their behaviors will change, and eventually alter the identities and the behaviors of a given nation-state in a self-perpetuating process of disintegration. As used in this study, "integration" will refer to the characteristics of a process of harmonization among Azerbaijan, Georgia, and Turkey that is developing, at least in the energy and transport spheres, supranational qualities which are distinct from the local qualities of each of the states' internal sectors. To this end, a process involving countries outside these three, or involving only two of the three, will not be treated as integrative for purposes of this study unless they produce at least a residual benefit for the third country. The point is simply that these three countries are engaged in an integrative process in the energy and transit sectors that unites the three of them to the exclusion of all other countries. The remainder of the study will seek to identify the causal mechanisms behind that process.

WHAT IS A TRANSNATIONAL EXTRA-REGIONAL ACTOR?

Transnational extra-regional actors (TERAs) are playing a significant role in the integrative processes among Azerbaijan, Georgia, and Turkey because they provide incentives in the form of financial donations and technical assistance to "cultivate" and encourage continued integration. The definition of a TERA includes two elements obvious from the name itself:

TERAs are external to Azerbaijan, Georgia, and Turkey, and they are multilateral institutions that possess organizational agendas different from those of a nation-state. In this sense, despite some similarities, TERAs are triggering an integrative process that is different from the hegemonic integration of an external nation-state.[21] They possess broad-based agendas of transparency, good governance, and the rule of law that are far more comprehensive than simple profit-seeking.[22] In other words, they stand to benefit from their involvement in the integrative process not because they are directly involved in the resource-extraction process, but rather because their involvement cultivates a broader shift in loyalties.

Like hegemonic integration, in which an external nation-state attempts to drive the integrative process for its own political and economic agenda, TERAs can spur integration by providing financial and other rewards, although they lack the coercive apparatus of a state. In other words, they can provide a carrot but not a stick. If, for instance, integration among Azerbaijan, Georgia, and Turkey was being spurred by the Russian Federation, a nation-state, rather than by the World Bank or the European Union, one would expect to see an integrative process based on Russian interests. This is not to say that, unlike a nation-state, TERAs necessarily act altruistically in their agendas; indeed, both the World Bank and the European Union are pursuing a certain economic agenda within their broad commitments to reform, one that favors foreign investment and a retreat of state involvement. Nor is this necessarily to suggest that the "extra-regional" nature of TERAs makes them qualitatively different from regional actors. However, the power of TERAs rests in large part on their access to global donor and technical-assistance networks that regional actors operating in Azerbaijan, Georgia, or Turkey may not be able to reach.

Two TERAs in particular will be analyzed for their involvement in the integration process in Azerbaijan, Georgia, and Turkey: the World Bank Group, for its involvement in the oil and gas sector, and two European Union–sponsored projects in the transport and energy sectors. The mechanisms by which the two programs work are slightly different. The World Bank is engaged in a broad-based agenda of economic development and good governance; by imposing strict conditions on the aid that it provides to recipient countries, it is able to achieve reforms in comprehensive fashion.[23] The European Union–sponsored projects TRACECA (Transportation Corridor Europe-Caucasus-Asia) and INOGATE (Interstate Oil and Gas Transport to Europe) are sector-specific projects that seek to harmonize the regulatory frameworks of Azerbaijan, Georgia, and Turkey and encourage their economic intertwinement with European Union member countries. In this sense, they operate less broadly but more deeply than the World Bank does, although, like the World Bank, the funds come with strings attached.[24]

All of the integrative theories described above have implications for the operations of TERAs in Azerbaijan, Georgia, and Turkey. When viewed from a security communities perspective, TERAs regularize and institutionalize contacts among the countries in which they work, improving access to information and reducing transaction costs. The result is that popular expectations change and loyalties begin to shift as people become more dependent on the newly harmonized system. This is arguably occurring in Azerbaijan, Georgia, and Turkey particularly with the transport sector, as rail, highway, port, and border infrastructure is improved and standardized. When viewed from a neofunctionalist perspective, TERAs use their technical expertise to engage in regulatory reform in order to improve the legal-rational nature of state bureaucratic governance. While this integration has not produced downward pressure for regulation in other spheres (spillover), such as how the harmonization of European steel and coal production placed pressure on railroad and transport integration in the early 1960s, it is true that TERAs are cultivating and accelerating their own continued development in a manner reminiscent of original neofunctionalism's determinism. Finally, TERAs fit within neoliberal institutionalism because they possess corporate agendas separate from the agendas of their member states; they interact directly with governments; and they create a framework for continued collaboration among the member states. Although Keohane wrote that countries would join international institutions to achieve national interests, TERAs by definition are extra-regional and thus do not necessarily include as members the states where they operate.[25] In the current case, however, the presence of TERAs in Azerbaijan, Georgia, and Turkey is based on consent and cooperation even without technical membership.

Ultimately, however, neoliberal institutionalism serves as a kind of unifying theory of integration in this study because TERAs are in large part driving regulatory harmonization, the neofunctionalist mechanism for integration, and shaping popular loyalties and expectations, the security communities mechanism. Neoliberal institutionalism, recognizing that both regulatory harmonization and the shaping of popular identities can be cultivated from the top down—and are not wholly organic processes working from the bottom up—is best able to explain the current integrative process in Azerbaijan, Georgia, and Turkey. However, elements of neofunctionalism and security communities theory provide neoliberal institutionalism with additional depth because they explicitly analyze two different mechanisms for integration that are especially important in the transport sector, which is a highly regulated, diffuse sector that interacts at many levels with the general population.

The integrative process currently taking place among Azerbaijan, Georgia, and Turkey is in sharp contrast to the historical isolation that separated the mountainous republics and produced dizzying diversity in language, religion, and cultural politics. Like other regions of the world that have received significant academic treatment, such as Western Europe, Southern Africa, or Southeast Asia, the integration of these three countries is helping to create an emerging if limited community of states that have little history of collaboration but that have found mutual benefit in cooperative arrangements. This subregion, however, possesses a distinctive pattern of integration that focuses almost entirely on the energy and transport sectors, which have several characteristics that may make them particularly amenable to the integrative process. First, they are heavily regulated sectors in which the state exercises dominant if not monopoly control, which has resulted in a state-centric harmonization process with a finite number of actors and strong compliance from participants. Second, both energy and transport are valuable enterprises that promise to bring financial reward as integration reduces transaction costs for both state-owned and private enterprises, who stand to benefit from the reductions of inefficiencies and barriers to export. Third, the energy and transport sectors possess a natural complementarity, with Azerbaijan having the advantage in resource extraction, Georgia possessing a political advantage as a transport corridor, and Turkey having deep ties to global markets and international financial institutions. Because of these unique characteristics, integration among Azerbaijan, Georgia, and Turkey has a distinctive pattern.

As noted above, integration has attracted significant scholarly commentary, spawning different theoretical frameworks, each of which emphasizes a different mechanism, or trigger, for why and how countries become increasingly interdependent. Chapter 1 of this study turns to the theoretical literature on integration in international relations theory, with an emphasis on the three primary comprehensive theories of how and why countries integrate. According to security communities theory, the first of the comprehensive international theories, the triggering mechanisms involve communications and transactions between populations that increase exposure, reduce transaction costs over time, and create a shared sense of common space that changes the communal identities of actors and ultimately alters state behavior. Neofunctionalism, by contrast, focuses on administrative and bureaucratic harmonization at low, technical levels, thereby placing regulatory pressure on larger and more significant sectors in a process called "spillover." Neoliberal institutionalism holds that states will become members of international institutions to pursue their national interests, but as the institution acquires its own corporate presence, it will alter the behavior of states by increasing the costs of defection and reducing the costs of compliance. This produces a top-down integrative process. While none of these theories can account for every case—neofunctionalism in particular has run into doctrinal pushback as it attempted to theorize beyond the Western European case—they each seek to fill a gap in the realist theory of international relations by explaining why countries engage in an integrative process even when the balance of power suggests that they should behave differently.

Integration among Azerbaijan, Georgia, and Turkey is a process by which certain national sectors are increasingly unified into a supranational system. Chapter 2 turns to the different state sectors in each country and assesses the extent to which the goals and expectations of each country overlap in order to determine the prospects for integration in a given sector. Cooperation in certain fields such as transport and energy suggests that the national interests of the three states are converging as a result of their cooperation, and this convergence of national interests is changing state behavior accordingly. To this definition it is necessary to add another layer of complication to help explain the Azerbaijan-Georgia-Turkey case. In these countries, integration is coordinated and cultivated in part by external actors, specifically referred to here as "transnational extra-regional actors," or TERAs, that provide incentives to integration in the form of financial and technical assistance to promote harmonization in the energy and transport sectors. These organizations, including the World Bank and the European Union, possess broad-based agendas of transparency, good governance, and rule of law that exceed in scope the profit-seeking agendas of foreign oil and gas companies simply seeking access to global markets.

In analyzing the Azerbaijan-Georgia-Turkey case, neoliberal institutionalism emerges as a controlling theory, particularly as to the energy sector. Chapter 3 is a case study of the World Bank Group's efforts in the energy sector in Azerbaijan and the development of an export route through the Caucasus—in particular, the pipeline from Baku, Azerbaijan, to Tbilisi, Georgia, and Ceyhan, Turkey. Because the energy sector in Azerbaijan is removed from the ordinary political system and controlled from the highest levels of government (and, geographically, because Azerbaijan's energy resources are clustered close to its center of power), international institutions such as TERAs are best able to access the decision-making process in this sector because of the financial and technical rewards that they can supply. Chapter 4, by contrast, focuses on the European Union projects TRACECA and INOGATE, which are harmonizing the transport sector among the three countries. Transport, unlike energy, is a more decentralized sector, although it possesses strong state involvement. Transport is a sector that is accessible to large segments of the population, allowing them to shape expectations about it. TERAs, in turn, place pressure on state and private enterprises alike to be transparent with accounting and profit sharing in both sectors. Because both energy and transport require assistance from neighbors—Georgia is the transport corridor for Azerbaijan's oil, while both seek Turkey's access to markets—TERAs reinforce the natural compatibility among the three countries. The two case studies relate to different international institutions and different domestic sectors; at their root, however, an integrative process is common to both.

In chapter 5, the analysis will turn to a counterfactual that possesses structural characteristics similar to the energy sector and pipeline network in Azerbaijan, Georgia, and Turkey: the natural gas export routes from Turkmenistan to China. Like the pipeline from Baku to Ceyhan, the natural gas export pipeline is a high-cost project that traverses several international boundaries among countries that do not possess a significant history of cooperation or openness. The natural gas pipeline, however, was directed by financing and assistance from the Chinese government and ancillary actors, including the state-owned energy company, rather than from international institutions such as TERAs. As a result, the secondary consequences from the presence of TERAs, such as transparency with oil revenues, objective and nondiscriminatory regulations, and formalized decision-making and dispute-resolution mechanisms, are absent in the Central Asian case. Although as in Azerbaijan the energy sectors in Turkmenistan, Kazakhstan, and Uzbekistan are highly state centric, they lack the institutional pressure for transparency and reform that the presence of TERAs provide to Azerbaijan.

Importantly, however, this pattern of integration in the energy and transport sectors has not led to integration in other spheres. The three countries still possess varying foreign policies, particularly as to Russia and the European Union, and on a security level Turkey is both unable and unwilling to defend Georgia and Azerbaijan from Russian encroachment. On an economic level, while some progress is evident from EU preferences as to all three countries and market reform in Georgia since the Rose Revolution, Azerbaijan, Georgia, and Turkey still possess economic agendas that are not entirely complementary. Likewise, in the social and cultural realms, linguistic, religious, and cultural orientations continue to divide the three countries. There has not been spillover in the neofunctional sense. Neoliberal institutionalism is better able to explain this segmentation: without the existence of TERAs operating in these spheres to provide financial and technical incentives to integration and to cultivate integration from above, then comprehensive harmonization along the lines of Western Europe will remain elusive. The following chapters will provide a comprehensive overview of the three primary modern theories of integration, compare and contrast their strengths and weaknesses, and assess their compatibility with and relevance to the case of Azerbaijan, Georgia, and Turkey.

NOTES

1. See generally Thomas De Waal, *The Caucasus: An Introduction* (Oxford: Oxford University Press, 2010).

2. Compare, e.g., Frank Möller, *Thinking Peaceful Change: Baltic Security Policies and Security Community Building* (Syracuse, NY: Syracuse University Press, 2007) and Amitav Acharya, *Constructing a Security Community in Southeast Asia: ASEAN and the Problem of Regional Order* (New York: Routledge, 2009).

3. See, e.g., Paul Stephenson, "Let's Get Physical: The European Commission and Cultivated Spillover in Completing the Single Market's Transport Infrastructure," *Journal of European Public Policy* 17, no. 7 (2010): 1039–57 (explaining cultivated integration from the top down in a neofunctionalist framework).

4. David I. Hoffman, "Azerbaijan: The Politicization of Oil," in *Energy and Conflict in Central Asia and the Caucasus*, ed. Robert Ebel and Rajan Menon (New York: Rowman & Littlefield, 2000), 55–78.

5. See, e.g., Stanley Hoffmann, "Obstinate or Obsolete? The Fate of the Nation-State and the Case of Western Europe," *Daedalus* 95, no. 3 (1966): 862–915.

6. Steve Smith, "International Theory and European Integration," in *International Relations Theory and the Politics of European Integration: Power, Security, and Community*, ed. Morten Kelstrup and Michael Charles Williams (London: Psychology Press, 2000), 34.

7. Dimitris N. Chryssochoou, *Theorizing European Integration* (London: Routledge, 2009), 15.

8. Karl W. Deutsch, *Political Community in the North Atlantic Area: International Organization in the Light of Historical Experience* (Princeton, NJ: Princeton University Press, 1957); see also Emanuel Adler and Michael Barnett, eds., *Security Communities* (New York: Cambridge University Press, 1998).

9. Ole Waever, "Insecurity, Security, and Asecurity in the West European Non-War Community," in *Security Communities*, ed. Emanuel Adler and Michael Barnett (New York: Cambridge University Press, 1998), 71–72, 104–107.

10. Emanuel Adler and Michael Barnett, "A Framework for the Study of Security Communities," in *Security Communities*, ed. Adler and Barnett (New York: Cambridge University Press, 1998), 44. According to Adler and Barnett, social learning plays a critical role in the emergence of security communities, and is facilitated by transactions that typically occur in organized settings. During their transactions and social exchanges, people communicate to each other their self-understandings, perceptions of reality, and normative expectations. As a result, changes occur in individual and collective understandings and values. Ibid.

11. Ernst B. Haas, *The Uniting of Europe: Political, Social, and Economic Forces, 1950–1957* (Notre Dame, IN: University of Notre Dame Press, 1968); Leon N. Lindberg, *The Political Dynamics of European Economic Integration* (Stanford, CA: Stanford University Press, 1963).

12. Arne Niemann and Philippe Schmitter, "Neofunctionalism," in *European Integration Theory*, ed. Antje Wiener and Thomas Diez (Oxford: Oxford University Press, 2009), 63 (describing neofunctionalism as partial theory). For the resurgence of neofunctionalism after the end of the Cold War due to the resurgence of European integration, see ibid., 45.

13. Robert O. Keohane, *After Hegemony: Cooperation and Discord in the World Political Economy* (Princeton, NJ: Princeton University Press, 2005).

14. See, e.g., Yelena Kalyuzhnova, "Overcoming the Curse of Hydrocarbon: Goals and Governance in the Oil Funds of Kazakhstan and Azerbaijan," *Comparative Economic Studies* 48, no. 3 (2006): 583–613.

15. Joseph S. Nye, "Multinational Enterprises and Prospects for Regional and Global Political Integration," *The Annals of the American Academy of Political and Social Science* 403 (1973): 117–19.

16. Ibid., 126.

17. Ibid., 118 (see table).

18. Jack Donnelly, *Realism and International Relations* (New York: Cambridge University Press, 2000), 9–10.

19. Michael Suhr, "Robert O. Keohane: A Contemporary Classic," in *The Future of International Relations: Masters in the Making*, ed. Iver Neumann (London: Psychology Press, 1997), 91.

20. Ibid., 92.

21. Above, I indicated that integration as defined in this thesis is distinguishable from coercive forms of amalgamation such as colonialism and military conquest. But simply because consent is necessary for the theory does not necessarily mean that integration cannot be controlled or initiated by a more powerful actor over a weaker one.

22. Howard Stein, *Beyond the World Bank Agenda: An Institutional Approach to Development* (University of Chicago Press, 2008), 43–44 (describing the expansion of the World Bank agenda from a state-oriented one focused on civil service and parastatal reform in the 1980s to a broad-based one involving governance).

23. The World Bank adds conditions not only relating to repayment, but also "conditions that the recipient country must fulfil by changing some of its previous policies." Paul Mosley, Jane Harrigan, and John Toye, *Aid and Power: The World Bank and Policy-Based Lending*, vol. 1, *Analysis and Policy Proposals* (New York: Routledge, 1995), xiii. For a more critical perspective, see Nancy Alexander, "Judge and Jury: The World Bank's Scorecard for Borrowing Governments," in *Fear and Want: Obstacles to Human Security*, Social Watch Report 2004 (Montevideo: Social Watch, 2004), 17–23.

24. Bernard Touboul, "The Central Asia Harmonisation of Border Crossing Procedures in the Perspective of TRACECA," paper presented at the Regional Seminar on Trade Facilitation and Customs Modernization, Issyk-Kul, Kyrgyzstan, August 4–8, 2003; Damien Helly and Giorgi Gogia, "Georgian Security and the Role of the West," in *Statehood and Security: Georgia After the Rose Revolution*, ed. Bruno Coppieters and Robert Legvold (Cambridge, MA: MIT Press, 2005), 270–306.

25. See Keohane, *After Hegemony*, x–xi. As noted above, however, the "extra-regional" characteristic of TERAs is not analytically necessary for the process described here.

List of Abbreviations

ACG: Azeri-bayu-Guneshli
ADB: Asian Development Bank
ASEAN: Association of Southeast Asian Nations
BEEPS: Business Environment and Enterprise Performance Surveys
BRIC: Brazil, Russia, India, and China
BSEC: Black Sea Economic Cooperation
BTC: Baku-Tbilisi-Ceyhan pipeline
BTE: Baku-Tbilisi-Erzurum pipeline
BTK: Baku-Tbilisi-Kars railway
CIP: Community Investment Programme
CIS: Commonwealth of Independent States
CNPC: Chinese National Petroleum Company
CPS: Country Partnership Strategy
CSTO: Collective Security Treaty Organization
DB: World Bank–IFC annual *Doing Business* report
DFID: Department for International Development
EAEC or Euratom: European Atomic Energy Community
EBRD: European Bank for Reconstruction and Development
ECOWAS: Economic Community of West African States
ECO: Economic Cooperation Organization
ECSC: European Coal and Steel Community
EEC: European Economic Community
EITI: Extractive Industry Transparency Initiative
ENP: European Neighborhood Policy
EU: European Union
FSAP: Financial Sector Assessment Program
GATT: General Agreement on Tariffs and Trade
GTZ: Deutsche Gesellschaft für Technische Zusammenarbit ("German Technical Cooperation")
GUAM: Georgia, Ukraine, Azerbaijan, Moldova
IBSU: International Black Sea University
IDA: International Development Association
IEG: Independent Evaluation Group
IFC: International Finance Corporation
IFI: International Financial Institution
IMF: International Monetary Fund
INOGATE: Interstate Oil and Gas Transport to Europe

JNA: Joint Needs Assessment
KfW: Kreditanstalt für Wiederaufbau ("Reconstruction Credit Institute")
LAFTA: Latin American Free Trade Association
MCC: Millennium Challenge Corporation
NATO: North Atlantic Treaty Organization
NGO: Non-Government Organization
NIS: Newly Independent States
OPEC: Organization of Petroleum Exporting Countries
OSCE: Organization for Security and Cooperation in Europe
PSA: Production-sharing agreement
RDI: Regional Development Initiative
SADC: Southern African Development Community
SCO: Shanghai Cooperation Organization
SFF: Supplier Finance Facility
SOCAR: State Oil Company of the Azerbaijan Republic
SOFAZ: State Oil Fund of Azerbaijan
TACIS: Technical Assistance to the Commonwealth of Independent States
TEN: Trans-European Networks
TERA: Trans-national extra regional actors
TRACECA: Transport Corridor Europe-Caucasus-Asia
UNDP: United Nations Development Programme
UNOMIG: United Nations Observer Mission in Georgia
USAID: United States Agency for International Development
USTDA: United States Trade and Development Agency
WTO: World Trade Organization

ONE

Toward a Theory of Integration

As noted earlier, integration is defined as the emergence of (a) at least rudimentary institutional structures among the countries involved; (b) interdependence in policy formation and implementation; and (c) a collective sense of mutual identity and obligation.[1] Each of the three theories of integration presented in this chapter focuses on different aspects of the integrative process as contained in this starting definition: communications or transactions among populations (security communities theory); intertwinement of government policies and regulatory frameworks (neofunctionalism); and facilitated harmonization by supranational agencies (neoliberal institutionalism). The purported goal of each of these theories is to explain what traditional realist theory cannot fully explain: namely, why former rivals have sought to cooperate, especially at an accelerated pace after the end of the Cold War, even when doing so appears to be at odds with national security and hard-power interests while eventually changing those interests over time. Unlike the traditional realist assumption that a state, acting in its own self-interest, exists in an anarchic world without a higher authority, each of these theories conceives of national interests as changeable and multifaceted, and of the authority structure of the international system as more complex than anarchy. Although the three theories emphasize different elements of the integrative process, and the permutations of hybrid theories include even more, the three theories have a common core by which the interests of states begin to converge as a result of their cooperation, eventually causing state behavior to change. This chapter will explore the origins and development of these three theories of integration and their continued relevance in the modern world.

To some degree, all three of the traditional integrative theories overlap, and each has borne an array of hybrid and descendant theories. Of particular relevance to this study, however, is that the theory of neoliberal institutionalism can be broadened to encompass both neofunctionalism's regulatory harmonization pressures and security communities theory's emphasis on popular communications and transactions. Neoliberal institutionalism in particular offers a framework for understanding how supranational organizations can cultivate or engineer regulatory transformation or reduce transaction costs for communications and other transborder contacts. This may be especially true among countries that have little common history of cooperation but possess strategically important advantages, such as the energy and transport sectors of Azerbaijan, Georgia, and Turkey, the case study under review here. Because these countries are developing at a rapid pace, they are particularly promising targets for international institutions that possess funding and technical assistance. The compatibility of the three traditional theories with neoliberal institutionalism makes the latter theory more promising as a unifying explanation for integration in the current case study.

SECURITY COMMUNITIES THEORY

Security communities theory developed during a time of massive reorientation of the field of international relations away from the classical realist doctrine that perceived of states as the only unique actors on the international scene. According to realist theory, sovereign states, recognizing no higher authority, exist in an anarchic framework that forces them to exist in a condition of mutual competition and conflict.[2] Although realist theory was undoubtedly dominant in international relations theory by the twentieth century, it was not the only paradigm. The most theoretically comprehensive challenge to realism was from the sixteenth-century thinker Hugo Grotius, who viewed the international sphere as a society of nations that possessed a common framework of moral and legal norms, and not as anarchic. During and after the 1950s, Grotian theory, always a minority viewpoint, enjoyed something of a renaissance. Deutsch's publication in 1957 of *Political Community and the North Atlantic Area: International Organization in Light of Historical Experience* was the most explicit advancement of the Grotian paradigm in this era.[3]

Deutsch's contribution to international relations theory is immense, if underappreciated. Underlying Deutsch's belief that violent conflict could be eliminated from international relations was his observation that there have always been pairs and clusters of countries who have had peaceful relations and whose leaders and peoples did not expect a resort to violence. Deutsch never really conceived of international relations strictly as interactions among states, but rather as a world of peoples, communities, or identity groupings, characterized by "within and without" and "we and they" perceptions and relationships. Degrees of mutual identification among peoples influence probabilities of peaceful conflict resolution among their governments.[4] This emphasis on identity and community foreshadowed the constructivist revolt against realism that blossomed in the 1990s. Today, much of security communities theory is underpinned to constructivist notions of social identity and interests.

Deutsch's Classical Security Communities Theory

As Deutsch and his colleagues set out in their initial study *Political Community and the North Atlantic Area,* political unions can be divided into two fundamentally distinct groups: amalgamated and pluralistic. Amalgamated political unions, such as the United States or unified Germany or Italy, are composed of formerly sovereign units into a single larger unit with a common government, while pluralistic political unions are composed of units that retain their political sovereignty. Both amalgamated and pluralistic political unions can form security communities, but do not necessarily always do so if the political units fail to integrate. Integration refers to the crossing of a threshold, Deutsch wrote, from a situation where war between political units was possible and anticipated to another situation where it was neither.[5] Integration involved a fairly broad zone of transition rather than a narrow threshold, and states could and did cross and recross that threshold several times, and could spend decades wavering. Using this definition of integration, an amalgamated political union such as the Austro-Hungarian Empire or a pluralistic union such as the Germanic Confederation would not have qualified as security communities because they simply did not integrate in a way that would prevent conflict among the component parts.

While amalgamated security communities, such as the United States after 1865, Great Britain after 1701, or the German or Italian states since the second half of the nineteenth century, undoubtedly have the potential to form highly well-integrated and stable political units, their failure is often bloody, as the US Civil War, the Irish independence struggle, the disintegration of Austria-Hungary, and other examples clearly illustrate.[6] By contrast, pluralistic security communities among states retaining their sovereignty almost never disintegrate to the point where the political components engage in hostilities toward one another.[7] Deutsch's team famously concluded, "[o]n balance, therefore, we found pluralistic security-communities to be a more promising approach to the elimination of war over large areas than we had thought at the onset of our inquiry."[8]

Essential requirements for amalgamated security communities include shared values and expectations among the communities concerned; unbroken links of social communication between political units and politically relevant elites; mobility of persons across political units; a fairly wide range of common administrative functions and state services; and mutual predictability of behavior among the member political units. Amalgamation, as Deutsch is careful to point out, is a *process*, one that accelerates as the political units benefit from closer union. Components that increased size, power, and economic strength during the amalgamation process undoubtedly contributed to further amalgamation. Increasing communication and other links furthers the process. In addition, amalgamated political units often developed around core areas—such as England (Great Britain), Prussia (Germany), Piedmont (Italy), Virginia/Pennsylvania/Massachusetts (United States)—that were larger, stronger, and more advanced in administration, political culture, and economic development. It was around these core units that the integrative process was centered.[9] But political amalgamation increases the demand upon the material resources and decision-making capabilities of the center, since decisions over a greater number of people and a wider area are made by a smaller number of institutions. Where claims and burdens on the central polity outstrip the polity's capabilities or political will, the amalgamation process will cease and reverse. The failure of the British Parliament to respond adequately to the Irish potato famine in 1846 resulted in a disintegrative process. Irish-English relations deteriorated between 1880 and 1914 even as the Irish became more "English" in matters of education, political rights, language, and economic class structure. Increasing links, in this instance, did not further political integration on the whole.

The background conditions and developmental process of pluralistic security communities do not differ greatly from amalgamated security communities. "Members of an amalgamated security-community—and, to a lesser extent, of a pluralistic security-community—must be able to expect from one another some dependable interlocking, interchanging, or at least compatible behavior; and they must therefore be able, at least to that extent, to predict one another's actions," Deutsch wrote.[10] At least three preconditions were necessary to the formation of a pluralistic security community, in common with amalgamated security communities: (1) compatibility of values relevant to political decision making; (2) capacity of the participating political units to respond to each other's needs, messages, and actions quickly and without resort to violence; and (3) mutual predictability of behavior. Other characteristics are helpful, such as a distinctive way of life, superior economic growth, broadening of elites—the free movement of labor between the United States and Canada during the nineteenth century in particular contributed to the formation of a pluralistic security community between them—but these considerations do not appear to be required. Again, the formation of a pluralistic security community should be conceived as a process rather than a state of being: although the existence of a broad range of social communications between communities can be seen as a background precondition to the existence of a pluralistic security community, the development of those communicative links and transactions is necessary to the formative process of the community.

The primary impetus to and characteristic of the formation of a pluralistic security community is the increasing unattractiveness and improbability of war among the political units concerned. This is particularly true where warfare would be devastating and indecisive, as, for instance, between the United States and Canada after 1815. While the land forces of the United States would likely have overwhelmed Canada, the naval power of Great Britain had the potential to devastate the US coast. In other cases, international complications threatened to engulf the contestants. Warfare between Norway and Sweden after 1905 could have involved Britain, Germany, or Russia; the formation of a security community between these two countries avoided that risk. Military alliances alone, however, are poor pathways to the formation of a security community; more is required in nonmilitary relationships. Similarly, foreign military threats were helpful to integration but not required. Undoubtedly, however, the existence of stronger and more capable political units, not weaker ones, contributes to stronger pluralistic security communities, and improving the range and volume of communication and transactions between units increases responsiveness to demands, which contributes to integrative forces. Like amalgamated security communities, pluralistic security communities also form around cores of well-integrated countries. The North Atlantic Treaty Organization (NATO), for instance, was preceded by smaller groups of well-integrated countries (the United States and Canada, the United Kingdom and the Netherlands). Crosscutting political coalitions and even broad-based mass support are increasingly essential for the existence of pluralistic security communities in an era of mass political participation. Stronger communications, however, remain key to the security community development process: communication not only allows greater exchange of goods, people, and ideas, but also permits more robust consultations between governments to make more unified decisions.

In his later works, Deutsch further developed his theory of integration. Integration, he wrote, was a relationship among political units in which they are mutually interdependent and jointly produce system properties that they separately lack. An integrated system is cohesive to the extent that it can withstand strain and resist disruptions. Cohesiveness among German or Italian states after unification is underscored by the fact that no component part has attempted to secede, even after devastating defeats in war; by contrast, Austria-Hungary and Great Britain–Ireland were much less cohesive under stress.[11] An amalgamated political community is more attractive if the common goals are something more than just peace, such as greater power, the acquisition of a common identity, or the fulfillment of some specific task. If the main goal is the preservation of peace, a pluralistic security community may suffice and be easier to attain. He also theorized the process of *disintegration*: a security community could unravel due to steep increases in economic, military, or political burdens; rapid increase in political mobilization or political participation faster than assimilation in the community's shared political culture; rapid increase in regional, economic, linguistic or ethnic differentiation, faster than any compensating integrative process; serious decline in political or administrative responsibilities of the center; the relative closure of the political elites, slowing the entry of new members and giving rise to counterelites; or the unstable loss of power by a formerly privileged or dominant minority, such as white settlers in the Federation of Rhodesia and Nyasaland in the 1950s.[12]

Functionalism came in for special criticism in Deutsch's later works.[13] Classical functionalism as theorized by David Mitrany is based on the hope that by delegating common technical and administrative tasks to supranational organizations, nations will grow to become administratively intertwined, making war more costly or impossible. As Deutsch notes, however, international administrative bodies are often limited in function and often cannot do their own binding rulemaking; they cannot tax, and most cannot sanction violators. Except for technical assistance, these bodies do not interact with the people they claim to serve. Popular loyalty to these agencies at the expense of national institutions is unlikely to grow since visibility is so low. The greater the potential power of an international administrative body, the greater the resistance will be of nation-states to the body's control. The World Bank and International Monetary Fund, which possess unusually broad monetary powers and can block access to credit, are subject to particularly strong criticism that will prevent further extension of their power. Functionalism essentially means partial amalgamation, as specific tasks are handed over to common agencies by participating political units. These tasks are often not very important and do not require additional amalgamation to be conducted efficiently. While some amalgamated security communities had functional amalgamation, such as the German Customs Union in the German states before unification, others did not: Italian unification, for instance, was not preceded by functional amalgamation, and functional amalgamation between Norway and Sweden could not ensure their merger. While functionalism has little effect on the eventual success or failure of amalgamated security communities, it can help participating governments learn habits and skills that could be useful in more far-reaching integration.

Deutsch has called integration an "assembly-line process" in the sense that the essential elements need to be present but do not need to develop in any particular order. The process usually begins with coalescence around relatively more highly developed political units. Early in the process a psychological "no-war" community develops in which war among the political units becomes illegitimate and lacks popular support, and where entangling alliances require conflict, hostilities are often kept to a minimum. This kind of relationship existed among the Swiss cantons in the sixteenth century and the Italian states in the eighteenth century; it also emerged among European Economic Community (EEC) members in the 1950s. The most salient political divisions within the amalgamated community become weaker and shift away from the boundaries of the political units. Political life then becomes divided by divisions that cut across the political units. The more varied and salient these cost-cutting ties are, the better for the cohesiveness of the union. Conversely, where cleavages among regions and political units are reinforced by old or new cleavages of language, social standing, or religion, integration may be halted or reversed, as between the North and South in the United States in the early nineteenth century.

Writing soon after Deutsch, Feldstein applied his theory to labor migration on the European continent, concluding that the flow of individuals as a transaction unit is part of a multidimensional process of political integration resulting in a transfer of loyalties.[14] While an influx of workers from outside the EEC could sharpen the division between the communal "we" and the foreign "they," the makeup of labor migration among the members of the EEC will shift as the economy improves, especially emigration from Italy. As labor migration moves away from low-skilled, low-status workers to a broader range of labor migrants, social communications increase between member countries as person-to-person contacts could become more regular.[15] As with any transaction or communication, merely the increased transaction flow of European workers does not necessarily spark interpersonal contacts or a "European spirit," because the migration is so fluid and dynamic. However, programs such as trainee exchanges that link labor migration to economic rewards and cooperation among EEC members could create meaningful transactions that make integration self-reinforcing.[16]

Refining Deutsch's Theory at the End of the Cold War

Deutsch never really conceived of international relations strictly as interactions among states; indeed, his writings consistently pictured a world of peoples, identity groupings, and communities shaping perceptions and relationships. According to Puchala, his most significant contributions to the field of international integration are his findings concerning sentimental relations among people.[17] Today, scholars view Deutsch not only as the original designer of security communities theory, but as a foundational voice in constructivist theory, which has been ascendant in the field of international relations since the mid-1990s. Unlike realist notions that anarchic relations among nation-states are a basic assumption of the international system, constructivists believe that anarchy is learned and ultimately becomes a self-fulfilling prophecy as states treat other states according to their expectations. Consequently, identities and patterns of state behavior can be changed over time.

Deutsch and his team anticipated many ingredients of the current social-constructivist strain of modern international relations theory, challenging the predominant realist framework in several ways.[18] While in the realist paradigm weaker states will seek to form alliances to counter the strength of a more powerful state, in the security communities paradigm weaker states are attracted to more powerful "core states." By focusing on beliefs rather than on material factors, Deutsch rejected realism's mechanical understanding of state behavior as objectively quantifiable, especially in terms of military capability. Finally, by positing that durable expectations of peaceful change could exist between states, Deutsch undermined a central element of realist assumptions about the security dilemma of nation-states. A community requires people to believe in and expect peaceful change, forming a sense of community or a "we-feeling."[19] Security communities are mental constructions underpinned by individual and collective experience and by shared knowledge. They are also normative constructions in the sense that the collective population affirmatively believes that intergroup relations based on peaceful change are "superior" to and more desirable than other forms of intergroup relations.[20]

As Adler and Barnett explain, security communities theory never generated a robust research agenda during the Cold War because Deutsch's conceptualization contained several theoretical and methodological problems that distracted from the essential theory.[21] Foremost among these was Deutsch's overemphasis on quantifying stages of integration through mechanical factors rather than on developing a more abstract security communities framework. Consequently, his focus on the volume and quantity of transactions between member states rather than on the social relations generated by the transactions ultimately constrained his work. He also failed to develop the roles of international organizations, crosscutting social classes, and political elites and charismatic individuals in the integrative process. Against the realist backdrop of the Cold War, a study focusing on security communities seemed overly romantic. His study was frequently cited but generally not emulated.[22]

Adler and Barnett's revival of Deutsch's theory with their work *Security Communities* in 1998 reinvigorated a concept that threatened to fade into obscurity.[23] Adler and Barnett made several important contributions of their own to the concept of security communities that corrected some of the problems with Deutsch's original theory. One weak point of Deutsch's theory was the slippery binary distinction between amalgamated and pluralistic security communities, which Adler and Barnett replaced with a spectrum. Security communities can be categorized according to their depth of trust, the level of institutionalization of the governance system, and the extent to which they have transformed their anarchic environment, Adler and Barnett determined. These factors provide a basis for distinguishing between two ideal types: loosely coupled security communities and tightly coupled security communities. Loosely coupled security communities observe the minimal definitional properties: a transnational region of sovereign states with dependable expectations of peaceful resolution of disputes. Tightly coupled security communities have a "mutual aid" system in which the units assist one another in a collective arrangement. They also possess a common governing regime that is postsovereign, endowed with common supranational institutions and some form of collective security.[24]

The next contribution that Adler and Barnett made to Deutsch's original theory was to develop the notion of mutual trust between political units as the linchpin of a security community. Here, Deutsch's omission of international organizations and multilateral nonsecurity communities becomes salient. International institutional arrangements, even those not built around security concerns, contribute to the development of mutual trust among community members by establishing norms of behavior, monitoring mechanisms, and sanctions to penalize violators. Because international institutions transmit shared expectations across borders and manage transactions between peoples, they contribute to shared learning among the societies of each political unit that help the integrative process. As shared expectations and self-understandings are communicated among societies, changes occur in individual and collective values. International institutions can transmit these expectations across borders and generations. While communication and transactions between people and society play a crucial role in the development of security communities, the growth of these communities is not organic; it requires political elites and charismatic individuals to turn the propensities into reality.[25] By explaining how nonstate actors such as international organizations and political elites can contribute to the formation of a security community, Adler and Barnett have compensated for Deutsch's overreliance on the state as sole actor in security community development.

Another way in which Adler and Barnett contributed to Deutsch's theory was by refining the phases of development of a security community in distinguishing between nascent, ascendant, and mature security communities. In the nascent phase, governments seek to increase mutual security, lower transaction costs, and encourage further exchanges. Political units must recognize or discover that they have joint interests that require collective action and can mutually benefit from some minimal coordination of security policies. This can happen due to an external security threat, due to a desire to capitalize on an international division of labor, or due to cultural, political, social, or ideological homogeneity. As Adler and Barnett explain, a security community "gets out of the gate" because of push-or-pull factors that cause states to reconsider how they organize their relations.[26] A more highly developed core state or group of states can provide leadership, protection, material benefits, and sense of purpose, as Deutsch recognized.

The ascendant and mature phases are further along the spectrum of developmental phases. The ascendant phase involves increasingly dense networks, the creation of new institutions, the decreased fear of threats posed by other members, deepening trust, and an emerging collective identity. Communication between populations increases knowledge that each has mutual conceptions of society, culture, and economics in mind, not just self-interested material gains. As these expectations are increasingly institutionalized in the region, war becomes improbable and a security community emerges. Here, the distinction between loosely coupled and tightly coupled security communities becomes relevant. In a mature loosely coupled security community, states identify with one another and proclaim a similar "way of life"; there are informal governance structures built around a collective identity; and there is an expectation that states will practice self-restraint even during confrontations. Characteristics of these communities include multilateral decision making, unfortified borders, changes in military planning, common definitions of external threats, and using a discourse of conciliation and community. In a mature tightly coupled security community, mutual aid becomes a matter of habit and national identity. Characteristics of these communities include cooperative and collective security, a high level of military integration, free movement of people, and policy coordination against threats.[27] Fundamentally, Adler and Barnett conclude that pluralistic communities are not simply precursors to amalgamated communities, as Deutsch implied. Rather, pluralistic security communities are semipermanent international actors that can be highly sophisticated, even more so than historical amalgamated communities, a radically new form of regional governance.

At least theoretically, one may wonder whether a community of regionally based security communities operating essentially as protostates is really less anarchic, and consequently less warlike, than a community of states. Alex Bellamy attempts to answer this theoretical objection in a 2004 book, *Security Communities and Their Neighbors: Regional Fortresses or Global Integrators*, which adds an additional layer of complexity to Deutsch's model. According to Bellamy, the anarchic state of international relations is not a natural state at all; rather, it is learned based on the expectations of states as to how they would be treated by other states. States will "learn" to act differently if they perceive their relationships with other states in terms of shared membership in a security community.[28] Adler and Barnett argued that member states will seek legitimacy for their actions from the community by justifying their actions in terms of the community's norms. This too was a weakness of Deutsch's original theory; he did not emphasize the changes in a *state's* behavior as a member of a security community. In mature security communities, the identity of the state itself is shaped by its membership in the security community. State behavior changes as a security community becomes more highly developed.

Just as liberal democracies may be more warlike toward nondemocracies even while they are more peaceful toward one another because they do not recognize the legitimacy of nondemocratic states, so too is it possible that security communities may take the functions of states and perceive outsiders as living in anarchic conditions characterized by the security dilemma.[29] Here, Bellamy distinguishes between regional fortresses, ambivalent communities, and integrationist communities. In regional fortresses, the boundaries between members and nonmembers are more rigid than those that existed before the formation of a security community, as members perceive outsiders as a threat to those norms and identities and act accordingly. In ambivalent communities, the relationships between members and nonmembers do not change after the formation of a security community, particularly when two states (one a member and one not a member) already have a well-defined relationship that is not affected by the security community. In integrationist communities, boundaries between insiders and outsiders grow softer. The logic of shared norms, common interests, and elite networks that exists within a security community projects outward without immediately incorporating outsiders into the community itself.[30]

This is the essential finding of Bellamy's research: the more mature and tightly coupled a security community is, the *less likely* it is to be a regional fortress.[31] Security communities do not simply represent a Huntingtonian nightmare of civilizations in perpetual conflict with one another. Rather, the process of security community formation actually changes the behavior of states away from a reliance on conflict. A more loosely coupled security community reproduces more realist security policies on its borders. This was true of early ASEAN states, in which the community as a whole viewed outsiders in the region with suspicion and even hostility. However, as ASEAN in particular has matured as a security community, nonmembers perceived the community as accomplishing its goals, making it more attractive. The more the region's leaders talked to each other in informal consultations, the more they realized they shared a common set of ideas. As a security community matures and integrates, realist language becomes less prevalent and membership in the community creates its own "common sense." The high integration of European states among each other spills over to their relations with nonmembers, and the internal patterns of conciliation and peaceful cooperation are also evident in relationships with outsiders. Similarly, as ASEAN has matured, the community as a whole has helped Vietnam reconcile its difficult relationship with China. Nascent loosely coupled security communities have much less in the way of independent common sense derived from their limited social interactions, so their members frame their worldviews more in realist terms than in more mature security communities.

In this way, more mature and more tightly coupled security communities do indeed mitigate and even eliminate violent conflict between states without geographical limits. The very fact that a security community is maturing and integrating implies that it is fulfilling the expectations of its members, making membership (or at least dialogue and cooperation) more attractive to outsiders. The changing worldview of the community's member units changes the language they use to articulate their interests and security politics. These new practices and discourses are observed by nonmembers, socializing nonmembers into the language and practices of the community. The security community and its neighbors learn not to be threatened by each other. The more tightly coupled a security community is, the more the boundaries between members and nonmembers become blurred.[32]

Like Bellamy, Möller explains how Deutsch's original theory does not necessarily lead to more peaceful relations overall, because security communities could develop hard boundaries that engage in anarchic, security-dilemma relations with nonmember states. The relations between members and nonmembers could resemble conventional relations among nation-states, with mistrust, security-dilemma politics, and the threat or use of force. This is especially true where security communities are geographically limited and the borders of the community are contiguous with nonmembers or with another security community. Möller identifies two characteristics that must be shared by security communities if the development of these communities is actually to lead to more peaceful relations overall: first, they require soft boundaries with nonmembers, which he calls "frontiers," rather than hard boundaries with military buildup; and second, they cannot maintain a standing collective army (which Deutsch's original theory contemplated) because it is the existence of these troops that nonmembers could find threatening.[33] Disarmament, starting with withdrawal of troops from common borders, is the ultimate proof of security members' commitment to peaceful change.

Linking Security Communities and Constructivist Theory

Constructivism is a comprehensive theory of international relations that developed in the mid-1990s and is based on many of the same assumptions upon which Karl Deutsch relied forty years before. Constructivist scholar Alexander Wendt challenges the core realist assumption that anarchy forces states into recurring security dilemmas. According to Wendt, whether a system is conflictual or peaceful is a function not of anarchy and power but of the shared culture created through discourse and social practices. Because an actor's self-identity is based on others' diplomatic gestures, states can reshape structure by process and can reconstitute their self-identity toward more peaceful relationships.[34] Wendt accepts realism's starting point: that nation-states are the primary actors in the international system and that states have a corporate identity and certain essential needs such as survival, autonomy, and economic well-being. Consequently, young states will tend to act in an egoist manner. However, only if a state continues conceiving of its own role in an egoist manner will the system continue. By casting other states in a nonegoist manner, states can begin to build collective identities that include other states in their definition of "self." Concrete interests are not simply given by the system; rather, socialized beliefs about what kinds of objectives are worth pursuing or avoiding will shape each state's actual interests.

Wendt famously stated that anarchy is what states made of it, and, subverting the realist paradigm, he identified three "cultures of anarchy" as ideal types in the international system. The first is "Hobbesian culture," which characterized relations among states until 1648, in which each state perceived of other states as "enemies" with no limits on the use of violence. In "Lockean culture," which has characterized international relations since 1948, states view each other as rivals but in principle accept the fact of one another's existence. Finally, in a "Kantian culture," which has emerged only in recent times, and only among liberal democracies, states play the role of friends, using force to settle disputes and working as a team against external security threats.[35]

At the same time, states can differ in the degree to which they internalize behavioral norms. In the first degree, consistent with neorealism, compliance with a behavioral norm is a function of coercion due to the power of other states. In the second degree, consistent with neoliberal institutionalism, actors conform to the norm not because they believe it is legitimate but because compliance is in their own self-interest. In the third degree, consistent with constructivism, states have internalized behavioral norms and identify with others' expectations, which are incorporated into the state's own identity. Given that there are three forms of culture depending on the norms internalized by the actors and three degrees of internalization of these norms, Wendt crosses the two axes and concludes that the culture in which states find themselves at any point in time depends on the social practices that reproduce or transform each actor's self-identity and perception of others. Because egoistic, violent mind-sets maintain egoistic, violent processes, a culture of realpolitik becomes self-reinforcing. If, over time, actors gesture differently and cast other actors in less self-centered manners, a Hobbesian culture can become a Lockean one, and a Lockean culture can become a Kantian one.[36]

Addressing the same question as Bellamy, namely whether security communities simply create their own definition of the anarchic security dilemma of nation-states, Möller uses Wendt's framework to explain the difference between anarchy among nation-states and "anarchy" among integrating security communities. "Deutschian anarchy," Möller writes, falls somewhere between the Lockean anarchy (states as rivals) and Kantian anarchy (democratic peace) and is composed of sovereign states that recognize and respect one another but do not bind each other. Unlike the Lockean paradigm, violence no longer occurs; unlike the Kantian paradigm, mutual aid or assistance is unnecessary. "Mature Deutschian anarchy" is essentially a détente among states without violence but also without greater commitment to collective norms or values. Moving beyond a Lockean rivalry among nation-states to a "Deutschian anarchy" is an important step toward banning the use of violence in international politics. Möller explains how this could explain how the relationship among Lithuania, Latvia, and Estonia and their neighbors Poland and Russia—states that do not have collective security arrangements among themselves, that still use the rhetoric of realism in their diplomatic relations, and that have few common values—can nonetheless remain relatively peaceful among themselves.

Möller has criticized Adler and Barnett's introduction of a values analysis into security communities theory, and has advocated a more conceptually pure definition of the theory.[37] According to Möller, Adler and Barnett deviated from the original theory by emphasizing compatibility of core values and a collective identity, while the original theory only emphasized vaguely compatible values and mutual responsiveness. In Deutsch's theory, members do not need to have any values in common, such as liberal democratic values, other than the belief in peaceful change. Nondemocracies can form security communities with democracies so long as they accept the principle of peaceful change and have compatible values that are capable of coexistence in a community with liberal democratic values. Likewise, Adler and Barnett are too ambitious when they require mutual aid of security community members. Security communities are not necessarily collective security arrangements, and a security community could exist in principle so long as the members are nonviolent toward one another even if they do not aid one another. Möller accepts Wendt's explanation that regional collective security arrangements and regional security communities can conceptually merge, as regional integration may require a commitment to mutual assistance in the case of aggression from outside the region.

While Möller has promoted a conceptually purer security communities paradigm that removes the normative and highly culturally contingent variables such as shared liberal democratic values, other scholars have promoted a definition of security communities theory that heavily emphasizes these values. Goodby, Buwalda, and Trenin, for instance, hold that security communities can disintegrate unless they develop a similar sense of identity or self-image, transparency (including military-to-military cooperation), and a reasonably healthy economy in the member units.[38] They refer to this form of security communities theory as leading to a "stable peace," and viewed the United States, the European

Union, and Russia as the obvious drivers of this stable peace. With this Möller would disagree in his use of a conceptually purer version of security communities theory. Groups of people may disagree on many major values in terms of their identities and interests, but at the same time still accept nonviolence and peaceful change as the basic principle in organizing mutual relations. Members of a security community do not need to share common core values so long as their values are not mutually exclusive of one another.[39]

Not all scholars accept a constructivist theory of integration. As Moravcsik writes, the constructivist paradigm may not be comprehensive enough to be empirically useful in determining whether integration is occurring or why.[40] Although constructivism as a metatheory holds that interests of a governmental elite reflect more general, deeper, collectively held ideas or discourses in the population and that these ideas or discourses only change at rare "critical junctures" such as times of crises, this does not easily translate into a testable theory in a social science sense. While constructivism may provide a metatheoretical basis for Deutsch's security communities integration theory, it may be an overstatement to say that constructivism has "subsumed" security communities theory.

The Relevance of Deutsch's Model in the Modern World

Cantori and Spiegel explain that "transactionalists" such as Deutsch concentrate on contacts, interchange, and communication between peoples and not international institutions or bureaucracies as indicators of successful integration. "In terms of the subject matter, the transactionalists—more than the neo-functionalists—have tended to concentrate on Western Europe," perhaps because of a lack of data about developing countries, as security communities scholarship has had a strong empirical component.[41] If this criticism was true in 1973, it is certainly not true today, and indeed security communities theory has shown itself remarkably resilient and adaptable to regions outside Europe.

One of the most comprehensive and sustained readings of Karl Deutsch using a non-European case study is Amitav Acharya's *Constructing a Security Community in Southeast Asia: ASEAN and the Problem of Regional Order* (2009). Integration theory, Acharya wrote, declined as the European Community faltered with respect to the Middle East oil crisis and American technological challenges in the 1970s. In addition, the liberal-pluralist explanation for integration proved to be a particularly Eurocentric flaw in the theory.[42] However, the orthodoxy of international relations theory, even more than the overemphasis on European integration, caused security communities theory to stagnate until the constructivist revolt against both neorealism and neoliberal institutionalism in the 1990s. Constructivism provides a comprehensive theoretical foundation, rooted in human psychology and in sociology, for Deutsch's focus on the development of collective perceptions and identifications among populations and the development of a "we-feeling." While norms only "regulate" behavior in neoliberal institutionalism, in constructivism (and in Deutsch) those norms can transform and redefine state behaviors.

When the Association of Southeast Asian Nations (ASEAN) was founded in 1967, the outlook was particularly grim, with many of the postcolonial members having weak sociopolitical cohesion, governmental legitimacy problems, interstate territorial disputes, ideological polarization and the exportation of Communist revolution, and external intervention. Nonetheless, by the 1990s the organization proved to be one of the most successful experiments in regional cooperation in the developing world, with none of the founding members ever engaging in hostilities against another member. While the Asian economic crisis in the late 1990s sparked criticism of ASEAN's effectiveness, the organization has produced a regional charter and has worked to improve relations with China and India. As Acharya notes, security communities theory provides a framework for interpreting ASEAN integration, viewing ASEAN regionalism as a process of interaction and socialization and examining regional identity formation.[43] In particular, neorealism and neoliberal institutionalism are insufficient to explain the ASEAN case because of the relative absence in the community of formal and legalistic methods of cooperation. Consequently, these other lenses will not be able to assess the extent to which a regional identity is forming in the absence of formal institutions and processes.[44] These emerging regional norms include noninterference in domestic affairs, pacific settlement of disputes, regional autonomy, aloofness from military pacts and a reluctance to deal with military or security issues, and important cultural norms, including an aversion to formal institutionalization and a preference for consensus among members.[45]

At the same time, Acharya's ASEAN analysis contributes to and expands security communities theory in several ways. Again linking security communities theory to constructivism, Acharya emphasizes the importance of institutions not in simply developing regulations that bind members, but in representing community norms and acting as agents of socialization, helping to develop and project a shared identity. Acharya emphasizes Deutsch's distinction between "no-war communities," in which members simply comply with regulatory norms banning conflict, and "security communities," in which the identity of the group encompasses peaceful cooperation and "we-feeling."[46] The construction of identity is central to Deutsch's "we-feeling," and plays a determinative role in state behavior, not a role derivative of material interests. While the founders of ASEAN initially had little conception of a regional identity, the organization has played a critical role in developing a sense of regional identity, a supranational "imagined community," in laying down the boundaries of the Southeast Asian region, and in drawing upon indigenous traditions of its own members and redefining principles of cooperation from the outside world.

Although domestic stability may be a necessary precursor to a security community (or at least, as Deutsch might say, the stabilization process must occur along with the integration process), democratic and liberal economic values likely are not a necessary condition. Looking at ASEAN's relationship with Myanmar, Kuhonta describes a security community founded not on liberal political traditions, but on political stability, regime security, and state sovereignty.[47] ASEAN's durable peace, dating from the organization's founding in 1967, has not benefited democratic reformers and movements in member states, even for emerging democratic forces in Myanmar. Deutsch did not identify liberal democratic values as necessary to the existence of a security community, indicating only that compatible political values were necessary for a durable bond among members. As Acharya indicates, regionalism in the developing world has tended to be state-centric, which invites opposition from international and domestic civil society groups on transparency issues.[48] Liberal democratic values are not required for the formation of a security community in the developing world, as the ASEAN case shows, but those values are not irrelevant: increased transparency, often a by-product of democratization, helps reduce mistrust and improve and regularize communications among members. Developing-world case studies, unlike European and North Atlantic case studies, may also provide some theoretical perspectives as to why security communities unravel, a point not stressed in Deutsch or Adler and Barnett. According to constructivist logic, security communities fail not due to shifting external threats or the balance of power, but rather due to the same changing identity factors that created them in the first place.[49]

Scholars have continued to problematize the notion that a security community is forming in Southeast Asia. As Busse writes, ASEAN's collective identity may be overstated by scholars.[50] The concept of Southeast Asia as a region and ASEAN as a subregion are uncritically accepted by scholars purely on the basis of geography even though the region was sharply divided by colonial powers until relatively recently. Instead of assuming the existence of regions based on their location on a map, regions should be studied as emerging from social practice and political interaction. No nation-state possesses a monolithic national interest; all of them are complex mosaics of ethnic and economic diversity. In the long run, ASEAN may need to alter its basic principles such as nonintervention and state-centric development, but the fact that it survived the Asian financial crisis without intraregional tension indicates that collective affinities for the organization outweighed the temptation to attack neighboring states. However, ASEAN's revered principle of noninterference was increasingly criticized by Thailand and the Philippines, the two most democratic states in the region. Democratization is likely to weaken some aspects of the "ASEAN Way."[51]

Besides ASEAN, the potential for the development of security communities on the African continent has attracted a considerable amount of scholarly attention recently. In part, the debate over whether regional security communities exist on the African continent is dependent on how one treats the historically high rate of internal conflict despite the relative absence of interstate conflict. According to Franke, the end of the Cold War marked a period of sharp transition in African politics, toppling revolutionary socialist states, American client states, and white-minority regimes alike, which led to a transformation of intra-African security cooperation.[52] The transition of the Organization of African Unity into the conceptually stronger African Union, as well as successful African interventions such as ECOWAS in Liberia in 1990, have helped shape continent-wide and regional frameworks for military intervention, conflict resolution, and security coordination among African countries.[53] More concretely, the African Stability Force and the Continental Early Warning System require sophisticated cooperation among national governments to prevent hostilities and contain them when they occur, and the adoption of a Non-Aggression and Common Defense Pact in 2005 gave the cooperation legal status in advance of African Union military interventions in Burundi, Darfur, Somalia, and the Comoros.[54] Rhetorical statements by many African leaders are now framed in the language of pan-Africanism rather than in realist terms. Looking at the Southern African Development Community (SADC) in particular, Franke concludes that a tightly coupled security community has formed in the region around the so-called Frontline States, which were committed to the end of minority rule in Africa and which closely collaborated on support for nationalist movements in South Africa, Namibia, and Zimbabwe. Moreover, the fact that many African countries are members of more than one collective security organization may also have the consequence of reducing conflict among them.

Franke's hypothesis that a security community is forming on the African continent, and in Southern Africa in particular, has generated a robust amount of commentary. Although Deutsch framed his security communities theory in terms of the absence of interstate war, Nathan has written that domestic stability must also be a necessary precondition to the existence of a security community. Drawing on examples from Central Africa and SADC, Nathan argues that domestic stability is a necessary precondition of a security community for three reasons.[55] First, given the stress Deutsch placed on communications and transactions *among populations* for the integrative process, a lack of personal security exhibited by gross violations of human rights, civil wars, or large-scale violence could preclude the existence of a security community because they may prevent these exchanges. Second, internal violence can and does lead to cross-border violence in the form of rebel attacks or military intervention, and at the very least increases suspicion among putative community members, reducing their mutual trust. Thirdly, domestic instability reduces consistency and predictability in state behavior, undermining another key element of Deutsch's security community preconditions. Nathan ultimately concludes that SADC or the countries of Central Africa could not be said to be a "nascent," "emerging," or "embryonic" security community, particularly given the unexpected hostilities shown among members in regard to the civil war in the Democratic Republic of the Congo.[56] Structural instability is characteristic of weak states; by contrast, Deutsch himself believed that strong states were more effective at establishing security communities. The absence of expectations for peaceful change among African populations belies elaborate formal mechanisms of security cooperation.[57]

Nathan is convincing when he questions the usefulness of a theory about the absence of interstate war on a continent where tens of millions of people have been victimized by active hostilities in the past fifty years. But Franke may also be correct that the existence of regional security mechanisms in Africa have, empirically, changed state behavior and have contributed to the development of regional and continent-wide identities. Deutsch's theory emphasized process, not end results, and it may be that the African context shows symptoms of a slow move toward a peaceful, cooperative end state. Regardless of whether SADC (or some combination of its members) or another regional organization is a nascent or emerging security community, Franke's suggestion that multilayered security cooperation among the African Union, SADC, ECOWAS in West Africa, the Community of Central African States, and the East African Community may contribute to the absence of interstate war on the African continent is an interesting and potentially useful theoretical distinction.[58] Unlike NATO-EU cooperation, for instance, African regional security communities must coordinate with each other (horizontally) and with the African Union (vertically) in an interlocking pattern, constraining the billiard balls in place.

The relationships of NATO and the European Union, both of which have some characteristics of "security communities," to countries on the periphery of Europe have attracted some scholarly attention. Moustakis, writing about Greek-Turkish relations that have historically been contentious until recently, describes Turkish integration into NATO as a "hybrid region," one possessing some but not all of the requirements for a fully integrated security community.[59] More recently, Moustakis and German have described the European Union's relationship with the Balkans and the Caucasus as efforts to create "zones of peace" and prevent security dilemmas from arising.[60] Not all cooperative efforts lead to peaceful outcomes, however, and the involvement by NATO and the European Union in these regions could actually lead to greater instability if external actors (namely Russia) view the security community as a threat. To that end, greater "integration" could undermine regional security.[61]

A Security Communities Theory of Integration

Integration in the security communities paradigm is the process by which communications and transactions among populations or political elites in different territories cause them to collectively identify to the point where active hostility becomes unlikely or impossible. Deutsch's original definition of integration was the broad zone of transition, from a situation where war was possible and was being prepared for, to another situation where it was neither.[62] Integration is a relationship among units in which they are mutually interdependent and jointly produce system properties which they separately lack. Political integration is the integration of political actors or political units in regard to their political behavior, such that the behavior of the units is modified from what it would otherwise be if the components were not integrated.[63] In order to qualify as a security community, groups of people have to expect and believe in peaceful change. They do so when a set of values perceived as compatible is translated into a sense of community, a "we-feeling." Security communities are to a large extent mental constructions underpinned by individual and collective experience as well as by shared knowledge.[64] This definition of integration through security communities is rooted in identities, interests, and expectations of populations rather than states as unified actors, and the mechanism of integration—aggregate communications and transactions between populations—involves the creation of web-like links between groups that lead to the formation of mutual trust and a common identity. This theory complements the modern constructivist revolt against realism because it views identity as mutable and socially constructed.

The value of security communities theory for a case study of integration is that it draws attention to nontraditional actors that realist theory does not reach. According to this theory, integration occurs not just when administrative linkages become more prominent or when governmental officials negotiate alliances, but in innumerable other ways: academic and scientific exchanges, travel and migration, or increased trade, for instance, all of which create points of contact and an increasingly solidifying sense of "we-ness" or community. By using a theory such as this to analyze political integration, a scholar is able to reach a broader scope of variables and actors to explain a complex and multidimensional process. Security communities theory has faced a revival since the end of the Cold War, and has relevance to continuing integrative processes all over the world, including in the postcolonial developing world.

NEOFUNCTIONALISM

Neofunctionalism was first formulated in the late 1950s and early 1960s, mainly through the work of Ernst Haas and Leon Lindberg in response to the establishment of the European Coal and Steel Community (ECSC), in which integration in the coal and steel sectors "spilled over" into much more comprehensive integration through the EEC and Euratom and again with the formation of a customs union ahead of schedule and a common agricultural policy.[65] The first edition of Ernst B. Haas's *The Uniting of Europe* (1958) is a dense empirical study of the early ECSC, but it also develops a set of general propositions about regional integration in the European experience. The theory declined in the 1960s, however, as European integration slowed through the efforts of French President Charles de Gaulle in paralyzing the governing commission of the EEC. Haas, Lindberg, and later scholars attempted several times to refine the theory, but it continued to decline until the close of the Cold War.

Neofunctionalism may be regarded as one particular offshoot of the liberal tradition in international relations theory, but it has developed into a sui generis theory, explaining in a rather ad hoc way the development of the European Union in terms of its own unique, sui generis characteristics rather than according to the terms of a general theory of world politics.[66] Neofunctionalism emphasized institutional form rather than socioeconomic content, with rational bureaucracy as a central driving force in integration.[67] Within ten years following Haas's *The Uniting of Europe*, the theory was forced to undergo significant revision, with marked changes made to the spillover principle that posited that pressure from integrated functional or political sectors would inevitably and automatically lead to integration in other sectors. Although neofunctionalism no longer operates as a comprehensive theory of political integration, certain aspects of the theory provide partial clues to integration processes outside the European context. Among these are neofunctionalism's emphasis on the role of interest groups, such as organized business and trade unions, in perpetuating the integration process, and the role of supranational elites of emerging international institutions that cultivate further integration.

The chapter will turn to the classical functionalism of David Mitrany, who pioneered the idea of a supranational technical agency that would administratively intertwine certain basic functions of the nation-state and assume legitimacy over time from the populations served by the supranational agency. Neofunctionalism, which arose in the 1950s, accepts the essential logic of functionalism but makes important deviations. The chapter will also look at neofunctionalism's attempts to reformulate itself as the process of European integration slowed down in the 1970s, as well as the ways in which the theory is of continuing relevance beyond the European integration project.

Classical Functionalism

The functionalist challenge to realism imagined a potentially cooperative welfare community in place of endemic anarchy. As O'Neill writes, an implicit assumption existed among functional theorists that cooperation in merely technical and commercial matters would have positive consequences for political integration, as illustrated by the expanding work of the League of Nations and the United Nations.[68] Writing at a time of gathering international crisis, Mitrany advocated global integration on the basis of administrative and technical cooperation, rather than political cooperation among nation-states or regional federations. According to classical functionalism, the only rational reply to the challenge that nationalism represents to world peace is the creation of supranational institutions that will gradually take over the function which the nation-state has traditionally performed but is no longer able to perform.[69] An international community must grow from the satisfaction of common needs shared by members of different nations. International agencies will develop a loyal community that supports the values of the organization. If such agencies were comprehensive and numerous enough, the loyalties to the international agencies and their communities would supersede loyalties to the nation-state.[70] The specialized agencies of the United Nations are far more comprehensive than those of the League of Nations and represent the first signs of loyalty to the world community. In delegating sovereign powers to international bodies, no power is more important than that of common security and defense, and the establishment of the Security Council and the International Atomic Energy Agency are the first signs that states are delegating their sovereign security power to international organizations.

Mitrany advocated functionalism, or the administrative intertwinement of nation-states through the delegation of certain discrete tasks to supranational bodies, over federalism, in which nation-states develop a global or regional supranational body while still retaining some sovereignty.[71] Because functionalism involves only discrete tasks such as interstate airline regulation, postal and telecommunications services, monetary policy, and other administrative functions, tricky questions of state equality do not arise in a functional system, as opposed to a federalist one.

The nation-state's inability to obtain common interests in interdependent conditions manifested itself as dysfunctional or incomplete, generating a demand for new institutional arrangements. Because those common interests could not be obtained with existing territorial divisions and national structures, those institutions had to be transcended.[72] Institutional form followed function, and the shape of the supranational institution depended on the function that it served. Functionalism explains institutional evolution in which common interests or needs are the initial motivation for the institution, but the institutional particulars are determined by environmental circumstances.[73]

Mitrany's theory has engendered multiple criticisms beyond simple naivety or idealism. Among these is doubt that the administrative tasks of the nation-state could be easily separated and dealt with more practically on a more global scale.[74] His belief that narrow technocratic solutions to complicated political problems can exist may not be a plausible reading of international politics.[75] Mitrany's attempt to depoliticize international relations confuses facts with values and ignores empirical evidence showing that international relations are, in large part, political.[76] Functionalism faces a social constructivist critique as well. The theory does not recognize the fact that the specific identities of nation-states shape their perceived interests and consequently their behavior in the international system. Functionalism assumes pregiven collective interests that lead to the formation of international institutions, and then assumes that new identities will form around those institutions.[77] However, the functionalist theory as conceived by Mitrany had certain durable qualities, including emphasis on decentralization, technocratic cooperation, and transnational administrative intertwinement, which became centerpieces of neofunctionalist theory. Neofunctionalism favored regional integration over global, and, unlike functionalism, possessed a scientific methodology and did not rely on claims of moralism.

One of the crucial links between functionalist thought and the theorists who formulated neofunctionalism were the proposals for sector integration put forth by Jean Monnet. The European Coal and Steel Community was Monnet's brainchild, framed as a modest surrender of sovereignty in exchange for strong economic advantages, but nonetheless part of a broader political agenda of European integration by stealth.[78] Monnet proved to be intimately involved with the European integration project despite several setbacks in the early 1950s and the only partial achievements brought about by the ECSC. By the mid-1950s, European integration focused more on removing barriers to trade and commercial cooperation rather than on positive sectoral and common-defense cooperation.

As Stefanou and Xanthaki write, Monnet was the link between Mitrany's functionalism and European integration, even though Monnet was not a functionalist per se and did not acknowledge Mitrany's influence.[79] Like basic functionalism, his plan involved the gradual transfer of loyalties and authority from the national level to the supranational level via the delegation of specific tasks to an international agency and its technocratic secretariat. However, the vision of Monnet and French Foreign Minister Robert Schuman was political, even if the means—pooling of coal and steel production—were economic, and their specifically intended goal of European political integration also shared characteristics with political federation, of which Mitrany so heavily disapproved.[80]

Monnet's approach to European unity was to gradually open the door to political federation by forging specific functional links among states in a way that did not directly challenge national sovereignty. While European integration did not proceed precisely in this manner, and, in fact, it may have proceeded much more rapidly as the functional approach declined in favor of state-centric efforts, Monnet certainly holds a symbolic role as a forefather of the European Union.[81]

In addition to his clear personal philosophy, Monnet possessed personal qualities that made him a highly effective proponent of European integration, and his American connections forged during World War II made him widely trusted among the Allied elites.[82] While coal and steel integration had been proposed many times as a tool for overcoming the historic antagonism between France and Germany, Monnet alone possessed the star power necessary to turn the Schuman Plan into the European Coal and Steel Community, his greatest accomplishment.[83] Although the Schuman Plan, proposed in 1950 and inaugurated as the ECSC in 1952, was not an economic or technocratic success, it set in motion the process that transformed Europe from a continent torn by nationalism to one economically and increasingly politically intertwined. The creation of a single, community-wide area in which coal and steel products could be sold competitively was a laudable but unrealistic goal, because coal and steel had been regulated for decades and each nation had its own system of control, obscuring real prices. Monnet knew that a true common market was impossible as the ECSC lacked the power to correct the root causes of market distortion.[84] When he resigned as president of the ECSC High Authority in 1954, the organization was beset by paralysis and its mission increasingly diluted due to a rise in prices during the Korean War, objections from technical experts, and competition from the oil industry.[85]

The ECSC did have technical successes: it regulated European trade in scrap metal, harmonized international freight rates, and launched the first major comparative study of taxation economics. However, the ECSC was even more successful for political reasons. The organization substituted as a peace treaty between France and Germany, creating the political underpinnings of a new international settlement that broke the impasse in Allied policy and allowed Germany to reenter Europe.[86] Although Monnet's plan for a European Defense Community failed and his vision for a customs union and for Euratom, the European Atomic Energy Agency, met with only lukewarm acceptance, the European integration project was relaunched in a round of diplomacy that culminated in the Treaty of Rome, signed in 1957, and the European Economic Community, formed in 1958. The EEC specifically rejected the ECSC model, but it likely would never have come into being had it not learned from the ECSC experience. Although Charles de Gaulle's return to power in 1958 ended the opportunity for EEC expansion, especially as to Britain and the

Nordic countries, de Gaulle did engage closer integration of the six original members of the EEC that increased trade among them sixfold and led to the dismantling of internal tariffs two years ahead of schedule.[87] With the difficulties faced by European and American currencies in the 1960s, European integration began to turn toward monetary union. Although these events did not directly descend from Monnet's quasi-functionalist vision and the ECSC experiment, Monnet helped change public and elite attitudes to integration that made the later efforts possible.

Ernst Haas and Neofunctionalist Theory

According to Haas, regional integration would occur when societal actors, in calculating their interests, decided to rely on supranational institutions rather than their own governments to realize their demands. Integration involves progress toward a consensus of the political community to impose limitations on intergroup conflict. This absence of violence within the political community is in some ways similar to Deutsch's security communities theory, but while Deutsch wrote that a specific institutional structure was not strictly required for his theory, as integration could take place among populations through transactions and communications, in Haas's neofunctionalism supranational institutions are a cornerstone of the theory. Haas's definition of political community as a population expressing loyalty toward central institutions is narrower than Deutsch's pluralistic security communities because central institutions are not required, but broader than Deutsch's amalgamated security communities because it could include intergovernmental agencies.

According to Haas's original theory, economic integration as it existed in the early European Community—characterized by elimination of tariffs, quotas, and exchange controls; harmonization of national policies to deal with the consequences; and freer movement of capital and labor—begat political consequences. As the beliefs and aspirations of groups change due to the necessity of working in a transnational institutional framework, mergers in values and doctrine come about, uniting groups across frontiers, implying the diminution of loyalty to and expectations from the formerly separate national governments. Political integration occurs when political actors in separate nation-states shift their loyalties and expectations to a new central authority, whose supranational institutions expand jurisdiction over national institutions. During initial stages of the process, loyalty to national institutions is still supreme, but as integration continues, a single political community emerges, superimposed over national political communities. The ECSC, a supranational organization, lies on a spectrum between two archetypes, federation and international organization. A federation divides sovereignty between the central authority and local governments, while an international organization treats only states as having legal obligations and rights. Federation decisions are binding while international organization decisions are not. International organizations are created by treaty and can be denounced; federations are intended to be permanent. Supranational institutions like the ECSC have some characteristics of political federations united by a single sovereign, but still have characteristics of international organizations. The ECSC possesses a federal executive, legislature, and judiciary but is compelled to act within the framework set by an international conference.[88]

Like functionalism, neofunctionalism emphasizes the mechanisms of technocratic decision making, incremental design, and learning processes, but beyond that the two theories diverge. Whereas functionalists believed that the administrative tasks assigned to an organization shaped the organization's mission, form, and scope, neofunctionalists placed great weight on organized interests, including civil society and political elites, in determining an organization's character. In addition, while functionalism was largely an analytical descriptive theory, neofunctionalists were sympathetic to the cause of European integration and intended their theory to be as prescriptive as it was descriptive.[89] Put in different words, although member states set the ground rules for supranational institutions, the direction in which those institutions develop is determined not only by member states but by two additional actors: (a) the secretariats and professional staff of the supranational institution itself, and (b) the interest groups, social movements, and associations that develop around the organization at a regional level.[90]

For Haas, realism's power-centered logic was as problematic as liberal idealism's pretense that a Kantian world order could transcend conflict. Haas agreed with Mitrany's belief that supranational institution building is premised on a technocratic engagement with human-welfare needs.[91] According to Haas, key social groups within nation-states would come to support deeper and more expansive integration as the integrative process proceeded, which he termed "spillover." As the ECSC example illustrated, an initial decision to integrate was likely to spawn pressures for deeper and wider integration, independently of any overt ideological preference for "more Europe."[92] Spillover was suggestive of automaticity—the idea that the logic of integration is self-sustaining, particularly as the integrated community itself acted as a strategic advocate on behalf of functional linkage and deeper and wider integration (which Haas termed "cultivated spillover"). However, as Rosamond explains, the concept of spillover unraveled in the 1960s as French President Charles de Gaulle led a reorganization of the European Community in an intergovernmental direction rather than a technocratic functional one.[93] Although spillover likely occurred in the European context, scholars generally agree that it has not occurred in other integrative contexts and is not germane to the definition of neofunctionalism.[94]

Haas's essential assumptions were these:[95]

1. Rational actors have the capacity to learn and change preferences, and as interests of actors change as a supranational institution grows in complexity and power, expectations and loyalties shift toward the institution.
2. Once established, institutions escape the control of their creators, and supranational elites encourage further integration by influencing perceptions of national elites.
3. Incrementalism, rather than grand designs, drives the integration process, particularly as many decisions are taken with imperfect information and under tight deadlines and thus produce unintended consequences.
4. Collective decision making emphasizes positive-sum games, instead of realist rhetoric promoting zero-sum games.
5. Emerging functional interdependence between the whole economies and their productive sectors inexorably tend to foster further integration.

Haas refined these points in a study with Philippe Schmitter about Mexico and integration in the Latin American context, which was a second-generation attempt to reach more universal variables leading to integration outside the European Community.[96] To Haas and Schmitter, the Western European case suggested that economic integration proceeded most rapidly and produced the most momentous political consequences when it accompanies a preexisting network of trade, travel, and communication as well as a transnational sense of loyalty and social mobility. Mexico's involvement with the Latin American Free Trade Association (LAFTA) appeared to be an aberration. Mexico's contacts ran north to the United States, rather than south to Latin America, which accounted for a miniscule fraction of its external trade, due to poor overland communications, long sea routes, and infrequent maritime service. In addition, Mexico possessed a strong indigenous culture and internal diversity. However, reasoned Haas and Schmitter, even though Mexico seemed to lack most, or any, of the preconditions that seemed to lead to integration in Western Europe, it possessed functional equivalents that drove political elites and business interests to a consensus in favor of regional integration.[97] In Mexico, that consensus was driven by a strong central state party that co-opted civil society and business interests and a dominant government bureaucracy of economic technocrats that could bind the state in ways that European technocrats could not. The existence of a strong administrative elite, articulating a widespread philosophy that economic integration would produce material wealth, engineered a top-down and state-centric consensus that made LAFTA membership very popular among the political elites. Thus, Mexico possessed an intergroup consensus in favor of integration even though it was constructed differently than in Western Europe.[98]

Leon Lindberg, who studied under Haas, introduced a conceptually purer theory of neofunctionalism with the publication of his book *The Political Dynamics of European Economic Integration* in 1963. Using Haas's definition of political integration as a process, Lindberg constructs a "more cautious conception of political integration," one in which collective decision-making procedures could achieve a significant amount of political integration without moving toward a "political community" per se.[99] To that end, Lindberg accepts political integration as a process but removes the reference to an end point. Lindberg introduces a two-part definition of the integrative process: first, nations forgo the desire to conduct policy independently and instead shift the decision-making process to new central organs; and second, political actors shift their expectations and political activities to the new center. While lacking some analytical clarity, Lindberg focuses on measurable expectations and political activities of elites rather than the more abstract values and loyalties around which Haas formulated his definition. Lindberg emphasized the process aspects of integration over the background conditions that allow integration to occur more than Haas did (and more than Deutsch in his security communities theory). According to Lindberg, for integration to occur, central institutions and policies must develop; the tasks assigned to those central institutions must be specific and important and of a type not normally performed by conventional international organizations; the tasks must be inherently expansive; and member states must continue to see their interests as consistent with the enterprise.[100]

Central institutions are required in order to represent the interests of and manage conflicts among the member states, and these institutions are well suited to uncovering sources of conflicts and breaking down the demands of states into more manageable components. The EEC, for instance, isolates issues and identifies common interests in order to promote unity. The tasks assigned to institutions such as the EEC must be inherently expansive and capable of overcoming what Haas called "the autonomy of functional contexts," in which low political functions are unbundled and become administratively autonomous.[101] This concept of spillover refers to a situation in which the original goal can be assured only by taking further actions, which in turn changes conditions and requires still more action. Lack of agreement among governments may encourage the growth of central institutions to which states can delegate difficult problems. For instance, backlash from nonmembers against a customs union may best be managed by further integration or expanding the role of the central authority.

Integration is rooted in the perception of actors that they can better satisfy their aspirations in the new institutional framework, Lindberg writes. What is so striking about the EEC, he writes, is the sheer breadth of the tasks assigned to the central institutions and the extent to which these tasks are inherently expansive as one functional context spills over into the next. An ever-widening circle of actors finds the system to be an effective, logical, and appropriate framework in which to pursue a state's goals.[102] The process of European integration could be reversed or halted despite the administrative spillover and intertwinement already achieved, but doing so would be procedurally costly for a state attempting to withdraw. Increasing membership in the Community to new states that are not as well integrated may be one means of slowing down the integrative process, but overpoliticization and reducing the institution to a technocracy could be other ways to halt the process.

The theory of neofunctionalism has several methodological problems beyond just empirical shortcomings with spillover theory. As Sterling-Folker writes, the continued reliance on functionalist assumptions about interests, institutions, and identity led neofunctionalists to overestimate the degree to which elite identities and loyalties could be changed by strategic interaction. Neofunctionalism assumes that elites are motivated by the desire to maximize their interest and would transfer loyalties and identities to the institution that provided the greatest rewards.[103]

Another methodological problem with early neofunctionalism was the disagreement among scholars over identifying dependent variables that could be quantitatively measured. Haas and Schmitter attempted to do this as early as 1964, by delineating nine independent variables. Nye criticized the variables because Haas and Schmitter did not explain the relationship among them, and their typology reads as a mere checklist with all factors to be weighed equally, even though there are strong reasons why some should be weighed more heavily.[104] Pointing to the East African countries of Kenya, Tanzania, and Uganda, Nye notes that political integration among the three had stagnated despite the fact that they possessed many of the Haas-Schmitter variables. The Haas-Schmitter framework also diverts attention from extraneous factors that cannot be fitted into the nine variables.[105] He advocates a theory that includes a "catalyst" or trigger to promote integration, beyond just recitation of a list of background conditions that favor integration. For his part, Haas recognized the symbolic nature of integration in the African and Latin American contexts.[106] In Africa, nation-state expectations are prematurely politicized, preventing cooperation even on relatively noncontroversial issues. In Latin America, size and power differentials inhibit economic union.

Schmitter revisited his and Haas's nine variables—categorized as background conditions prior to integration, conditions extant at the initiation of the integrative process, and process conditions once integration has started—in an article from 1969.[107] From this, he deduced three hypotheses about how integration proceeds. The first is a spillover hypothesis, in which members of an integration scheme agree on collective goals but are not satisfied with their attainment of these goals and thus attempt to resolve this dissatisfaction through either collaboration in another, related sector (expanding *scope* of commitment) or by intensifying commitment to the original sector (expanding *level* of commitment). *Scope* has two dimensions, the number of social groups or policy sectors involved and the importance of these policy actors for attainment of national goals. *Level* has two dimensions, the continuity of members in addressing their mutual obligations and the techniques of the policymaking process. The second hypothesis is an externalization hypothesis, in which participants find themselves compelled to formulate a common policy toward nonmembers. This is especially true if the bloc borders on another integrating bloc of nations. The third hypothesis is a politicization hypothesis, in which the scope and actions of the collective or supranational actor become increasingly controversial, which leads to a widening of the audience or clientele interested and active in integration and a redefinition of interests and eventually a shift in loyalties toward the supranational authority.

Lindberg attempted to solve this in 1971 by formulating a multivariable theory that focused on the breadth and content of the collective decision making in an integrating system, the level of responsiveness between collective decision makers and national counterparts, and compliance with and distributive consequences of the collective decision making.[108] In particular, Lindberg identified three categories of variables in his definition: (1) the level of collective decision-making activity, including the scope of policy areas affected, the range of stages involved in the decision process, and the degree to which collective decisions are the most decisive ones; (2) a collective decision-making body's response to the flow of demands to it from relevant political actors, determined by the level of resources available to the decision makers, the way in which leaders make use of them, and the propensity for national officials to "consume" resources for the maximization of collective interest; and (3) the consequences that collective decisions have for the constituent systems, including the degrees of penetration and compliance and distributive consequences. Such a complex set of variables is required, according to Lindberg, since earlier theoreticians focused on only one or a few of the variables, taking those variables to be indicative of the entire integrative process without explaining how the variables interacted with each other.[109]

The three layers of variables, as Lindberg indicates, exist on spectra and can change over time. For instance, the scope of collective decision making may be generalized over many issues or be specialized to a few policy domains. The range of participation in joint decision making may be limited to predecisional stages and information sharing, or it may involve comprehensive administration and enforcement. The extent to which change occurs in national systems depends on how deeply collective decision-making processes penetrate into constituent systems. In some cases, collective decisions will be only binding on government negotiators and only during negotiations, while in other cases collective decisions will be binding on national governments but enforcement left to national authorities, and in still other cases, collective decisions directly enforceable on individuals. The amount of change attributable to collective decisions will vary with national authorities' compliance with the decisions. The most salient factor, however, may be how the benefits of collective decision making are distributed among component national units, as the willingness to proceed with further integration is dependent on some balance of returns among the participating countries so that none seem disproportionately disadvantaged.[110]

The Decline of Spillover Theory

Neofunctionalism is probably the most heavily criticized integration theory. While many criticisms are based on a misreading of the doctrine, one consistent criticism is that neofunctionalism cannot provide a general theory of integration in all regional contexts, as it overemphasizes qualities characteristic of relatively developed economies and functioning democracies such as bureaucratic sophistication. The portion of the theory that appears to be most highly contingent, and consequently most criticized by later scholars, was Haas's belief in the automaticity of the spillover process, which caused both Haas and Lindberg to walk back from automatic spillover in later writings at the expense of some of the theory's theoretical consistency.[111]

Generally speaking, traditional "spillover" is shorthand for the process in which integration of one sector leads to technical pressures that push states toward integration in other sectors due to the interdependence of administrative functions such that it is impossible to isolate them from the rest. Haas identified, for instance, the pressure that integration of coal and steel production placed on European transportation networks to deliver the products to markets, encouraging countries to cooperate more closely in secondary sectors. This would be termed "functional spillover." The spillover process occurred not only in functional sectors, but also in bureaucratic interdependence as EU working groups, for instance, worked closely with officials from national bureaucracies and civil society, labor, and business organizations worked increasingly across national boundaries on issues of broader interest. This process is termed "political spillover." Lindberg, with his stress on human agency, believed that supranational institutions intentionally drive the spillover process—what he termed "cultivated spillover"—in order to augment their own power and scope of action.[112]

Integration theory requires a theory of disintegration, according to Schmitter.[113] The doctrine of spillover is no exception, and scholars have used a colorful set of words such as "spillback," "muddle-about," and "spill-around" to describe the phenomenon of disintegration. One way in which the integration process could be halted or reversed is through encapsulation, according to which, in times of a crisis in integration, national governments impose a collective intergovernmental solution at the expense of the supranational community.[114] Spillback, or a reverse of spillover, can occur when leaders (such as Charles de Gaulle or Margaret Thatcher) or interest groups, lobbies, or opposition parties, or even public opinion, are wary of further integration and can stall the process. Slowness in the EU enlargement process, such as opposition to Turkey's accession to membership, is emblematic of the spillback process.[115] With falling growth rates and rising unemployment in the 1970s, the European integration process slowed down, and efforts to broaden functional cooperation among states faced increasing skepticism. Spillover began to look like a fair-weather concept, something that simply occurred in favorable economic conditions.[116]

Perhaps the most important condition of spillover is that for the process to occur a structural pressure must be perceived as compelling since structures need human agents to turn structural pressures into evolving policy. In that sense, spillover is not automatic in a mechanical sense; it requires human agreement. Where the rationale for expanding functional cooperation and the support of political elites are both strong, spillover is likely to occur.[117] In this way, for instance, the open movement of labor across borders in Europe placed pressure on national home and justice ministries to coordinate immigration and asylum policies.

Cultural spillover also figures into early neofunctionalist thought, and Haas (but not Lindberg) identified values and ideas as variables that lead to and promote integration. Over time, so the theory went, cultural changes would occur in integrating European societies that would culminate in the refocusing of popular loyalties away from traditional national attachments to a new European identity. This theory, however, lacked methodological rigor and the link between values and behavior was never clearly explained.[118]

Schmitter has written that later scholars misunderstood the concept of automaticity in the spillover process. Automaticity did not assume that the process was linear or consistent, nor did it assume that the process was entirely unconscious, involuntary, or even effortless. On the contrary, neofunctionalist logic emphasized precisely the opposite: the role of bureaucratic practice, organizational ideology, and creative interventions by administrative and political elites. Functional interdependence is not enough for integration; the interdependence must be perceived and translated into expressions of interest and viable decision-making strategies.[119]

The ultimate problem with spillover theory is not that it does not necessarily define a genuine phenomenon—empirically, "spillover" can be used to describe the administrative intertwinement of a number of sectors in the European Community and elsewhere—but that it is not connected to a deeper integration theory and does not seem to be a condition of or a consequence of regional integration. Consequently, scholars have used the language of "spillover" to describe the growth of the supranational European transport network, energy regulation, and immigration policy, among others, but the theoretical use of these concepts is limited. Stephenson linked the efforts of the European Commission to prioritize fourteen transportation infrastructure projects in Europe in 1994 as a form of cultivated spillover, leading to nearly continent-wide integration proposals for ports, airports, satellites, motorways, and railways over the next decade.[120] Integration of European transportation networks since the 1980s, one of the later sectors to become integrated, was a top-down affair with heavy involvement by the Commission itself in controlling information, mediating differences of opinion, and presenting broad-based policy proposals directly to member governments. The concept of cultivated spillover helps to describe this type of integration where it occurs, but case studies such as this do not provide a theoretical explanation for why it occurs.

Revived Neofunctionalism as Partial Theory

In his 1975 book *The Obsolescence of Regional Integration Theory*, Haas argued that regional integration theory in the European context had become obsolete.[121] European integration was a "halfway house" in which both the nation-state and the European Community proved surprisingly durable. The authority of the Community is quite strong, and very stable in the original issue areas formed by government consensus in the 1950s. The legitimacy of the Community, however, is increasingly questioned as the organization has failed to reach agreements on high-politics issues that are high on national agendas. Whether issues are high-politics or low-politics issues depends on the business cycle—military affairs swaps places with research and development on the continental agenda—due to the absence of a monolithic nation-state or a dominating supranational community that can control the agenda.[122] The result is disjointed incrementalism and fragmented issue linkage. This "halfway house" of European integration cannot last, Haas wrote, as continuing attitudes toward and expectations of the Community shift as the continental agenda changes. Integration theories are becoming obsolete because they are inspired by a sense of orderly process to describe a process of disjointed incrementalism. This inspiration is not wrong, he wrote, but merely incomplete.[123]

The relaunching of the European integration process in the 1980s, and the corresponding revival of integration theory, rekindled interest in neofunctionalism. Contemporary commentators acknowledge the partial relevance of spillover and provide modest but ambiguous analyses of regional integration that are very different from the positivism of the early writings of Haas and his colleagues.[124] A number of hybrid or syncretic writings use elements of neofunctionalism in broader theoretical frameworks to explain integration, but this is true for security communities theory and neoliberal institutionalism as well. Neofunctionalism certainly survives in mainstream integrative thought, but no serious commentators defend the original neofunctionalist logic. By the 1970s, more state-centric, realist-based theories of integration, particularly intergovernmentalism, were rising to prominence as European integration slowed, but these too had their own theoretical limitations.

Given the difficulties neofunctionalists have had in developing a comprehensive and theoretically consistent set of principles, despite regular attempts to reformulate the theory, neofunctionalism may be best viewed as a partial theory, as Niemann and Schmitter write.[125] Indeed, hybrid theories have appeared regularly in scholarly literature, using elements of neofunctionalist thought combined with elements of other integration theories. For instance, Sweet and Sandholtz found Deutsch's theory of social exchange, communications, and transactions compatible with Haas's theory of the development of cross-national elites and interest groups and cultivated integration by the supranational entity itself.[126] Their hybrid theory of integration is a product of rule-making and policy coordination by central authorities that build on crosscutting relationships among elites and form their own center of gravity that becomes its own impetus toward further integration.

One way in which the concept of spillover may be of continuing relevance is in the realm of what Niemann and Schmitter call "geographic spillover." Neofunctionalism provides some explanation as to why a number of nonmember countries applied for European Union membership: the high level of integration made EU accession attractive and exclusion costly. The demand for EU enlargement can be derived from neofunctionalism's basic tenet: integration leads to tensions and demands that can only be resolved by taking further integrative action (in this case, geographical reach rather than sector expansion).[127] In addition, both the EU as an entity and the interest groups involved in EU policy making have promoted expansion to Central and Eastern Europe. Neofunctionalists predicted that political elites and interest groups would favor further integration (here, geographic enlargement) once they became aware of the benefits of existing integration. That appears to be happening in the European Union case at an abstract level. Member states have to rely on the EU to formulate a policy toward nonmembers, further expanding the power and reach of the organization and making it still more attractive to outsiders.

This concept of geographic spillover has been applied to Turkey's admission to the European Union. As MacMillan explains, scholars such as Haas and Schmitter recognized that increasing regional integration would lead to expansion of the community to include nonmembers.[128]

As the regional organization enlarges, its political weight increases, and the economic consequences of nonmembership and discrimination from the community also rise. The prospect of enlargement will tend to create functional pressures for increased integration, and political and nongovernmental elites will form coalitions in favor of enlargement. In addition, enlargement as a policy was cultivated by the supranational institution itself. MacMillan notes that in the Turkish case, Turkey's membership in the European customs union since the mid-1990s required harmonization of standards and legislation, an example of functional integration. With Turkish accession, the EU will have new borders in the Middle East and Caucasus, providing foreign policy challenges and opportunities for the EU as a whole. In addition, political integration occurred as elites in Turkey passed packages of Western-style law reforms to meet EU guidelines. In this sense, the concept of geographic spillover is a lens through which European expansion could be measured in neofunctionalist terms.

Another way in which neofunctionalism has struggled to remain relevant in international relations is the ambiguous relationship the theory has with constructivism, one of the dominant and ascendant paradigms of international relations. As Niemann has written, elements of neofunctionalism are compatible with constructivism, such as the foundational principle that actors in the international system are capable of "learning" during the integrative process as the values, loyalties, and expectations of political elites change and ultimately change actor behavior.[129]

Perhaps Haas's self-critique is the most cogent, characterizing neofunctionalism as a lens rather than a comprehensive theory.[130] Whereas security communities theory stresses the volume and rate of transactions or aggregate data of public opinion or contacts, neofunctionalists observe bargaining styles and strategies and case studies of growth and decay. Neither theory has been predictive in a positive way, but neither has been disproven either. Both have been better at predicting failures than successes—why integration failed rather than why it succeeded. Neofunctionalism never identified a terminal end goal to the integrative process or an ideal type, and consequently has been better at successfully predicting specific cases of spillover and bargaining styles than in achieving closure and predicting formation of a community as such.[131]

One key assumption of early neofunctionalism is that integration cannot start with the political, but rather begins on low-key economic and technical planes.[132] Integration must start with low politics, in other words, rather than high politics. This division originated with Stanley Hoffmann, who argued that technical and low-level cooperation by technocratic and nonpolitical elites could lead to increasingly relevant cooperation at higher political levels.[133] Like security communities theory, neofunctionalism has faced criticism for its neglect of high politics.[134] Indeed neofunctionalism's failure to predict the challenge posed to European integration by Charles de Gaulle was emblematic of this failure to account for high-political changes. In any event, the constructivist argument that issues can change priority depending on ever-changing state and societal interests has made the distinction between high and low politics even more theoretically slippery than it once may have been.

Toward a Synthesis

Scholars who have developed hybrid theories often draw out three aspects of neofunctionalism that are both compatible with other theories and empirically helpful in understanding integration in a given case. The first of these is neofunctionalism's interest in certain nonstate actors such as transnational business associations and trade unions, who benefit from reducing barriers on the movement of capital and labor, rather than neoliberal institutionalism's emphasis on states as primary actors and Deutsch's emphasis on populations and political elites. The second is the emphasis on technocratic cooperation on relatively noncontroversial issues; while administrative intertwinement does not necessarily imply automatic spillover to other sectors, cooperation in certain sectors could provide a blueprint for cooperation in other sectors. Finally, the neofunctional belief that supranational institutions develop their own centers of gravity and cultivate further integration is a relatively durable aspect that has survived the original conceptions of the theory. Some or all of these aspects of neofunctionalism have been found compatible with other theories and have produced a staggering number of hybrids.[135]

When neofunctionalism was revived in the 1990s, scholars did not reintroduce the determinism (in the form of automatic spillover) associated with the original theory, but they did emphasize the importance of functional spillover in limited contexts and the role of supranational organizations and interest groups. Since the 1990s, a large variety of hybrid theories have laid claim to some aspect of neofunctionalist thought, although they do not combine actors and integration mechanisms in the same way. New approaches stressed the multilayered nature of the policymaking process; trends and structures of domestic politics; transaction costs in the integration process; development and identification of national interests; and national state executives and bargaining elites.[136]

As Verdun explains, no single theory best accounts for European monetary integration, for example, and taking aspects from several different theories could best describe the process as a whole.[137] To the extent that changes in the global economy, such as competition from the United States and Pacific basin, motivated European countries to develop a single currency, theories stressing intergovernmental cooperation and national policy agendas rather than neofunctionalism are most relevant. Likewise, intergovernmentalist theories are best suited to explain how member states supported economic and monetary union as a way to make unpopular welfare and public-finance reforms that would otherwise have been politically costly. However, neofunctionalist thought, unlike government-driven theories of integration, can explain how the presence of societal groups made economic and monetary union easier and faster by providing additional legitimacy to national governments' surrender of sovereignty to the supranational level. The support of trade union confederations and national and continent-wide employers' associations very actively promoted the European objective. Economic and monetary union was possible from a neofunctional perspective, because it involved highly technical matters better coordinated at the supranational level rather than national levels, and from an intergovernmental perspective, because the states involved had converging policy objectives that made union beneficial. This example illustrates how neofunctionalist thought, in conjunction with other theories, may still make a contribution to integration literature.[138]

A Neofunctional Theory of Integration

In neofunctionalist terms, integration is the process of transferring exclusive expectations of benefits from the nation-state to some larger entity through the process of administrative and bureaucratic intertwinement of government policies and structures. It encompasses the process by virtue of which national actors of all sorts—including government officials, interest-group spokesmen, politicians, and ordinary people—cease to identify themselves and their future welfare entirely within their own national government and its policies.[139] This is not to confuse the process with the end result. A political union exists when the politicized decision-making process acquires legitimacy and authority. "Integration," on the other hand, is only the analytical term to describe evidence of increasing politization (i.e., formation of a polity), of shifting expectations, and of adaptation by actors to a new process of mutual accommodation. It need not be consistent or unidirectional.[140] The theory shares several important similarities with security communities integration theory, namely the emphasis on identity and loyalty shifts among a population. The mechanism, however, is different: instead of web-like transactional and communications linkages, integration in neofunctional terms involves technocratic and administrative intertwinement that becomes more complex and sophisticated over time.

Although neofunctionalism as a theory is no longer defensible in its original terms due to its empirical inability to explain the course of European integration and the difficulty of transferring the theory outside the European context, as a partial theory it is useful because it emphasizes variables that the other theories do not. Two basic qualities of neofunctionalism have appeared repeatedly in recent scholarly literature. The first is the emphasis on the development of transnational elites that would encourage integration, such as business and labor interests who profit from open access to markets and workers. This is an important distinction from state-centric theories of integration. The second is neofunctionalism's recognition that the supranational technocratic elites who depend on the emerging institution cultivate continued integration through, for example, coordinating national government officials to conform to the supranational body's agenda. Even if spillover per se is not consistent or automatic, the motivations of supranational elites are expansionist as they seek greater power and promote the supranational identity. Although neofunctionalism today is a complementary lens through which to view integrative processes rather than a comprehensive theory, it continues to possess some relevance for the study of integration, not only in Europe, but elsewhere as well.

NEOLIBERAL INSTITUTIONALISM

A third type of integration theory argues that states form collective institutions in order to achieve their own interests, but, once built, these institutions can alter state behavior by establishing and enforcing norms, by sharing information, and by regularizing contacts and communications. Neoliberal institutionalism describes a pattern whereby states construct cooperative relationships around centers of mutual interests along institutional lines. While neoliberal institutionalism became more prominent in the 1990s, the best representation of its core principles is *After Hegemony* by Robert O. Keohane (1984). Accepting as fact the existence of interdependence between states, Keohane notes the potential for strife and the problem for international order that this creates. The existence of institutions—defined by Keohane as "persistent and connected sets of rules (formal and informal) that prescribe behavioral roles, constrain activity, and shape expectations"[141]—can explain not only the development of customary and repetitive practices among states, but broader cooperative behavior. Because of complex interdependence among states, potential cooperation among states existed more often than the realist paradigm would suggest. Since the survival of most states is not constantly at risk, states are more interested in maximizing gains rather than in preventing gains by other states. Consequently, opportunities for collaboration exist beyond what a realist paradigm would posit by focusing instead on a state's relative gains.

After World War II, realism emerged as the dominant paradigm among international relations scholars, where it encountered the idealist challenge from Mitrany and later the neofunctionalism of Haas, which spawned numerous studies of European regional integration. During the 1970s and 1980s, the debate between realism and liberal institutionalism began to take more definite shape. Three especially important works, explored further below, appeared during this time that laid out many of the fundamental tenets of neoliberal institutionalism before Keohane's *After Hegemony*.[142] The first of these was a special issue of *International Organization* in 1971 on the topic of "Transnational Relations and World Politics," which challenged realist assumptions of state centricity in international relations by looking at various transnational nonstate actors such as the Catholic Church, the Ford Foundation, and multinational business, each of which had some impact on state behavior. The second book was Keohane and Nye's *Power and Interdependence* in 1977, which explicitly challenged realism with respect to the state-as-actor assumption, the relative importance of military security on foreign policy agendas, and the role of military force in international politics, among other issues. The third work was a special issue of *International Organization* in 1982, edited by Stephen Krasner, on the topic of "International Regimes," which set forth both realist and liberal views on international cooperation and institutions.[143]

Perhaps more than the other theories analyzed here, neoliberal institutionalism developed in part in a back-and-forth debate with neorealism to define the points of disagreement and to clarify essential aspects of each respective theory. The most seminal works in this exchange are summarized below. The initial section outlines Keohane's theory of neoliberal institutionalism, followed by a more nuanced and parsimonious theory of international institutions themselves and the different roles they could play. The neorealist critique of neoliberal institutionalism follows, as do Keohane's efforts to refine his theory and respond to some of the criticisms. Finally, this chapter concludes with a neoliberal institutionalist definition of integration that could be of general application.

The Theory of Neoliberal Institutionalism

Unlike security communities theory or neofunctionalism, neoliberal institutionalism as delineated by Keohane is formulated on a realist foundation rather than a constructivist one. That is, the growth of international institutions occurs not because of affinities, communications, or shifting loyalties among populations, but because institutionalization is in states' interests as it reduces transaction costs and encourages sharing of information. States do not cooperate out of altruism, empathy, or loyalty to the international project—as could be assumed by security communities theory or neofunctionalism—but rather because states want to advance the wealth and security of their own people and appropriate for themselves the power to do so.[144] States build international regimes in order to promote mutually beneficial cooperation, such as in reducing transaction costs for states, alleviating problems of asymmetrical information, and limiting uncertainty among members. At the same time, the existence of international regimes can change state policies through changing the costs and benefits of various alternatives. Institutionalization does not replace state self-interest; it complements it.

Realism provides a taut logical structure and pessimistic assumptions that prevent wishful thinking. However, the theory of realism needs revision. Unlike realist paradigms, Keohane did not perceive of the international order as constructed solely around the distribution of power. Rather, the distribution of information is also a variable, and international institutions can provide information, increase credibility, and build mutual trust between states. International systems that have institutions that generate a great deal of high-quality information for major actors are more cooperative than systems that do not have these institutions, even if the state interests are comparable in both.[145] This is the reason that international institutions have proliferated and strengthened after the end of the Cold War, despite realist predictions that institutions would wither with the end of superpower conflict. In short, the existence of international organizations is still in state self-interest even though their existence is not as necessary in an anarchic distribution-of-power sense because of the information-sharing and trust-building nature of institutions. Cooperation can even take place without hegemony, as American aloofness from the Kyoto Protocol and the International Criminal Court attest.[146] Regimes do not need to be universal to be successful; indeed, they can be especially successful when a small number of like-minded countries both make rules and enforce them.[147]

Traditional realist theory of international relations can explain alliance cooperation in the context of a balance of power but cannot explain system-wide patterns of cooperation that benefit many countries without being tied to an alliance system. During the 1950s, both realists and functionalists could interpret the proliferation of international institutions in the context of their own views; realists pointed to American hegemony while functionalists pointed to the desire for a rules-based world system. The rise of Europe's economic power during the 1960s after recovery from war, according to realists, caused international organizations to wane as American hegemony was challenged. In the modern world, it is unlikely that a hegemon will appear, either the United States or another country, and consequently realism should predict the end of international institutions. Indeed, Keohane highlighted realist thinker John Mearsheimer's prediction that NATO would not survive the end of the Cold War because of the removal of the Soviet threat to Europe as emblematic of a larger theoretical flaw in realist doctrine.[148] To the contrary, NATO survived and expanded geographically and functionally after the Cold War. States invest heavily in institutions, and the broad generalizations of realist theory do not successfully explain why.

Interdependence in the global political economy generates conflict. To limit discord, cooperation is necessary to align government policies to prevent economic shock or political turmoil for other governments.

While the existence of a hegemonic power can drive this cooperation, the proliferation of nuclear weapons in the modern world (which distributes power among nonhegemonic states) will ensure that cooperation in the future will be nonhegemonic. Although states are driven by their self-interest rather than their devotion to a common good, world politics is not a state of perpetual war, and states do have complementary interests. As Keohane explains, international institutions create the conditions for orderly multilateral negotiations, legitimate and delegitimate different types of state action, and facilitate linkages among issues within regimes and between regimes. They increase symmetry and improve the quality of information governments receive. They help to bring governments into continuing interaction with one another, reducing incentives to cheat and enhancing the value of reputation.[149]

As in neofunctionalism, neoliberal institutionalism requires that states delegate authority to supranational organizations that will later become capable of acting against the subsequent preferences of governments. Even more than that, however, institutions help states reach a collectively superior outcome by reducing the transaction costs of further negotiations and by providing the necessary information for reducing state uncertainty about the future behavior and preferences of other states.[150] States, through supranational organizations, establish rules for the distribution of gains, reduce the costs of coordinating their activities, monitor the behavior of other states, and mutually sanction noncompliance. The structure of an international institution depends in part on the severity of distributional conflict, enforcement problems, and uncertainty about the preferences of other actors.

The theory of neoliberal institutionalism uses a definition of "international regime" similar to that instituted by political scientist John Ruggie, which defined regimes as sets of "mutual expectations, rules and regulations, plans, organizational energies and financial commitments, which have been accepted by a group of states."[151] Krasner used a slightly different definition when he and his colleagues began to lay the foundations for neoliberal institutionalism: "sets of implicit or explicit principles, norms, rules, and decisionmaking procedures around which actors' expectations converge in a given area of international relations."[152] Neoliberal scholars did not question the definition of anarchy as the absence of an institutional hierarchy capable of enforcing rules in world politics; they questioned the frequent implication that in anarchy, institutions based on rules were absent. Chiefly by providing information to actors—and not necessarily by enforcing rules in a centralized manner—institutions could assist states to achieve their own objectives more efficiently, such as by monitoring compliance and enabling retaliation against cheaters.

Keohane's theory has been criticized on several grounds. To Keohane himself, the theory has an unavoidable weakness because it treats all states as generic units, without taking into account variations in domestic politics among them. Rather than treating states in terms of the populations that inhabit them, as Deutsch does, Keohane treats them as billiard balls. As he notes, however, there simply was no way to incorporate theories of domestic politics in a cogent and parsimonious way. He also admits his overoptimism at the time of the original publication of his book in 1984. Far from necessarily benefiting all nations, especially the least developed ones, organizations such as the World Trade Organization and GATT (especially the extension to intellectual property rights envisioned at the Uruguay round) have pursued an agenda that is most favorable to wealthy, highly developed states. This does not mean that cooperation is not desirable, even though some players have reduced bargaining power relative to other players. However, it does suggest a moral imperative for institutional systems to reallocate more resources to poorer countries and direct those resources toward some of the world's least advantaged people.

The empirical research program of neoliberal institutionalism remained limited to a handful of successful Western institutions such as NATO, World Trade Organization/GATT, or the European Union, at the expense of other institutions such as the Warsaw Pact or the Organization for Security and Cooperation in Europe.[153] As Hemmer and Katzenstein have written, pure formulations of both liberalism and realism are less convincing than hybrid theories that also incorporate important insights from constructivist theory. To liberalism, constructivism contributes an understanding of the consequences of personal and group identity to international institutions. To neorealism, constructivism adds consideration of ideational and non-material structures and the effects of identity on national interests.[154] Neither pure liberalism nor pure neorealism can explain why NATO has not been replicated in Asia. Constructivist theory suggests that the Asian continent lacked some components of a common identity that NATO countries shared. For neoliberal institutionalism, culture matters as to how information is distributed and interpreted. From a realist perspective, it is inexplicable that a strong state would join a common institution with weaker states and thereby bind its own future courses of action, as Germany did in supporting economic and monetary union.[155] NATO is viewed by its neighbors as a concrete manifestation of Western civilization, and consequently enlargement that was controversial at the time, such as Italy's inclusion, now seems natural. The Asia-Pacific region also lacks a history of institutionalization compared to Western Europe and North America. In this way, some aspects of neoliberal institutionalism may be compatible with aspects of constructivism.

At the same time, syncretism between neoliberal institutionalism and neofunctionalism is possible in large part because the two theories share basic presumptions and indeed common theoretical origins. Neoliberal institutionalism rests on a functionalist logic that assumes pursuit of and maximization of profit was a primary motivating factor for nation-states.[156] Neoliberal institutionalists, like functionalists, assume that institutional development would rest on the ability to deliver on collective goals in given environmental circumstances, and as circumstances changed, so too would institutions. According to neoliberal institutionalism, environmental circumstances were constantly evolving, and even diffuse institutional patterns could be explained with this same functional logic. The difference is that neoliberal institutionalism works within the nation-state paradigm rather than opposed to it, so that states could be partners in a process that, rather than replacing the state, would help it meet its own interests. Both theories view the state as an obstacle to work around, but neoliberal institutionalism retains a place for the state.

Developing a Theory of International Institutions

State actions depend to a considerable degree on the flow of information and opportunities to negotiate and the ability of governments to monitor others' compliance and to implement their own commitments and expectations about others' commitments. With these considerations in mind, international institutions can take one of three forms: first, formal intergovernmental or international nongovernmental arrangements; second, international regimes, or, in Oran Young's terms, "negotiated orders" such as the Bretton Woods system and the Law of the Sea regime; and third, conventions, which are informal institutions with implicit rules and understandings.[157] All three types overlap. According to Keohane, the level of institutional sophistication can be measured along three dimensions: first, commonality, or the degree to which expectations about appropriate behavior and understandings are shared by participants in the system; second, specificity, or the degree to which those expectations are clearly detailed in terms of rules; and third, autonomy, or the extent to which institutions can alter their own rules rather than relying on participating members to do so.[158]

Another theoretical ancestor of neoliberal institutionalism was the interdependence theory of the 1970s. In 1977, Keohane and Nye anticipated the theoretical background for a theory that ascribes significant influence to international organizations. According to Keohane and Nye, interdependence is created by reciprocal efforts resulting in mutual dependence among countries or actors in different countries, resulting from international transactions across boundaries.[159] Governments and transnational actors will strive to increase their share of the gains from these transactions, even where they both profit in absolute terms. Less dependent actors can often use the interdependent relationship as a source of bargaining power because changes in the relationship will be less costly to that member. In a world in which security issues are dominant and war threatens frequently, international institutions will play a relatively minor role. However, in a world where multiple issues are imperfectly linked, in which coalitions are formed transnationally and internationally, the potential bargaining role of international organizations is greatly increased. They can help set the agenda, act as catalysts for coalition building, and provide an arena for initiatives by weak states.[160] International organizations, by bringing government agencies and officials together, can create explicit coalitions founded on direct constitutions. In this way, a world characterized by complex interdependence looks different from a realist world.

In an earlier book, Keohane and Nye elaborated on the role that transnational communications (and international institutions as facilitators of such communications) play in the international system by linking groups in different societies and promoting attitude changes.[161] Here, these organizations are defined broadly to include not only intergovernmental organizations but also transnational civil society organizations and multinational corporations. These international institutions can create new symbols, myths, and norms to provide legitimacy for their activities. The very existence of international institutions and other organizations operating across borders implicates technological changes in communications, travel, and transportation, producing new connections that have formed the basis for neofunctionalism and security communities theory. Another major effect of the internationalization of institutions is the creation of dependence and interdependence. States may become dependent on international scientific, transportation, or financial networks, or even on international organizations that provide certain goods and services, which may alter domestic patterns of political relations to make otherwise feasible policies more costly, such as setting an autonomous monetary policy. Among unequal states, transnational organizations and networks may put additional means of leverage into the hands of more powerful states to the disadvantage of those that are already weak.[162] In short, transnational relations provide different sets of incentives or payoffs for states, because they alter the choices open to statesmen by affecting the relative costs of different policies. Transnational organizations have become large and powerful actors capable of consciously attempting to shape the world to their interests.[163]

In a later collection of essays, Keohane elaborates on the nature of international institutions.[164] Institutions have both constitutive and regulatory aspects: they affect the understandings of state leaders about their own interests and the interests of others, and at the same time they affect state behavior, even if evasion is possible. However, international social structures and organizations defined broadly exert much less pull on state policy than domestic institutions. Keohane distinguished his version of neoliberal institutionalism from liberal theory, indicating that cooperation, even among democracies, is not automatic; rather, it requires planning, negotiation, and the exercise of influence. Neoliberal institutionalists accept a version of liberal principles that avoids determinism and that emphasizes the pervasive significance of international institutions without denigrating the role of state power.[165] International regimes, and the institutions and procedures that develop in conjunction with them, perform the function of reducing uncertainty and risk by linking issues and improving the quantity and the quality of the information available to participants. Linking issues is important because it prevents deception or cheating, which is less profitable in an institutionalized relationship where the cheater's behavior is closely monitored by other participants and interaction among them is continuing. Institutions improve communications among participants also, but they cannot make it perfect: knowledge among bargaining parties is asymmetrical and government officials rarely fully reveal their preferences in international diplomacy.[166]

Considering these qualities of international institutions, it is possible to make some distinctions among different types of international regimes. International policy coordination depends not only on interests and power, but also on expectations and information, which are functions of domestic political structures and elites. Intergovernmental relationships characterized by ongoing communication among working-level officials are inherently more conducive to informational exchange than are traditional transnational relationships controlled by members of a government bureaucracy. Regimes with highly regularized procedures and rules will provide more information to participants than less regularized regimes; where information is in demand, these regimes will be more highly valued, as will regimes characterized by open government arrangements.[167] Creating international regimes creates risk that another state will fail to carry out its commitments, but they mitigate the consequences arising from rapid and unpredictable changes in world politics. Keohane also distinguishes between "control-oriented regimes," in which member states create institutionalized arrangements to monitor each other's behavior, and "insurance regimes," in which member states pool resources and risk, such as oil-sharing regimes.[168] While control-oriented regimes seek to control externalities, insurance regimes spread risk around when externalities cannot be controlled. Because control-oriented regimes require more stringent membership conditions, whether a member state chooses one form over another depends on how the benefits of regular, institutionalized contact weigh against the organizational and autonomy costs of submitting to the institution's rules. The optimal size of a regime will increase if there are increasing rather than diminishing returns to regime scale or if the marginal costs of organization decline as the regime size grows. In dense policy areas, demands for overall frameworks of rules to cover clusters of issues (and consequently, demand for international regimes) will increase.[169]

Martin has expanded this typology even further.[170] Different cooperation problems, she writes, lead to different institutional structures. She uses game theory to distinguish among four types of problems: collaboration (where the equilibrium outcome is bad for both parties); coordination (where each actor has a different preferred outcome); suasion (where a more powerful member persuades a weaker one to adopt a less favored outcome); and assurance (where both sides seek to benefit from cooperation). In collaboration problems with many actors, high incentives to engage in undetected cheating should trigger the emergence of a strong institution in the absence of a hegemon who can enforce compliance. In these types of regimes, "free riders" will need to face some sort of punishment in order for the regime to work (as with the World Trade Organization). Coordination problems, on the other hand, do not present significant advantages to concealed defection from established conventions, so multilateral norms may be sufficient to establish cooperation. In any event, institutions established at the early phases of the dispute may be quite successful. In suasion problems, cooperation is achieved through issue linkages, and an institution could work if it binds the dominant power and persuades smaller powers to join. Suasion problems, unlike the others, involve the core problem of asymmetry. In assurance problems, the institution will work to increase transparency, as the exchange of information is the crux of the contentious issue.[171]

As noted above, reciprocity between states requires the ability to recognize and retaliate against defectors. One way to do this is to establish a hierarchy whereby more powerful states, which are more committed to the regime, could deal effectively with weaker states who are in danger of defecting. These types of hierarchies exist, for instance, where international banking networks could facilitate rescheduling of developing-world debt. Another way to do this is through establishing international regimes, which have become ubiquitous after World War II in, for instance, the international trade regime (GATT/World Trade Organization) or the international monetary regime (International Monetary Fund and partners). Since the use of power could be channeled into either establishing a hierarchy or establishing a regime, the two forms of conditional cooperation should be seen as complementary. Regimes, including international institutions, do not enforce rules in a hierarchical sense, but they do change patterns of transaction costs and provide information to participants so that uncertainty is reduced.[172]

International regimes do not substitute for reciprocity between states; rather, they reinforce and institutionalize it. Regimes incorporating the norm of reciprocity make defection illegitimate and therefore more costly. Because they define reciprocity in the given issue area, they make it easier to establish a reputation for practicing reciprocity consistently. Such reputations are themselves important assets, because other states will be more willing to make agreements with governments that could be expected to respond to cooperation with cooperation. Since governments with good reputations can more easily make agreements than governments with bad ones, international regimes can help to facilitate cooperation by making it both easier and more desirable to acquire a good reputation.[173] In this sense, international institutions possess a kind of creditworthiness function. Occasionally, international regimes have been used to attempt to develop new norms, but they have not always been successful, such as international financial institutions with norms of lending and repayment. Establishing hierarchies or institutional regimes changes the context within which actors operate by changing the very structure of their interaction. They have been experimental, trial-and-error efforts to improve the current situation, and they have occasionally failed.[174]

The Neorealist Critique of Neoliberal Institutionalism

Neoliberal institutionalism appeared in the early 1980s after international tensions during the 1970s caused other liberal international relations theories to decline. As the realist scholar Greico explains, neoliberal institutionalism accepts a number of core realist propositions, including an assumption that systemic anarchy impedes cooperation, while remaining faithful to the liberal belief that international institutions promote cooperation and reduce conflict.[175] Greico surmised that neoliberal institutionalism misconstrued the nature of international anarchy and consequently, the theory's optimism about international cooperation would be proven wrong.[176] According to Greico's analysis of neoliberal institutionalism, states seek to maximize their absolute gains, and international institutions can improve compliance so that all states can benefit. However, realism also posits that states may not cooperate even where cooperation would be beneficial for all parties, if one state benefits more than another state; both absolute and relative gains can determine whether cooperation succeeds or fails. Unlike in neoliberal institutionalism, in neorealism states must give serious attention to the gains made by other states because they are potential enemies, a major source of state inhibitions toward cooperation that neoliberal institutionalism is unable to identify.

As Baldwin explains, six focal points characterize the debate between neorealism and neoliberalism: (1) the nature and consequences of anarchy, which have different conceptions in the two theories; (2) the ease and likelihood of cooperation, which neoliberal institutionalists believed contained unrealized potential; (3) relative versus absolute gains, with neorealists tending to focus on the former and neoliberals on the latter; (4) the priority of state economic versus state national security goals, with neoliberals emphasizing the former and neorealists the latter; (5) an emphasis on interests, intentions, and information (neoliberalism) versus distribution of capabilities (neorealism); and (6) the importance of institutions and international regimes.[177] Keohane did acknowledge in a 1993 essay that neoliberal institutionalists underestimated the importance of relative gains under certain conditions, but he emphasized the need to identify what those conditions are, since empirically it is difficult to tell when states are pursuing absolute or relative gains.[178]

Although neoliberal institutionalism claims to accept the premise of realist anarchy, neoliberal anarchy is characterized by the absence of a supranational body that could enact rules of behavior and force states to cooperate. In other words, states may wish to cooperate so that they all may benefit, but they lack a central agency to coordinate their expectations. For realists, on the other hand, states are not driven by a desire for well-being, but rather a desire for survival, and consequently states view other states' gains with suspicion. States are driven to maintain their position in the system more than they are driven to maximize their own power and well-being.[179] A state may avoid cooperation with other states if it believes its partners are achieving relatively greater gains, even if cooperation would have won it large absolute gains. A state may even forgo cooperation even if other states do not cheat, transaction costs are low, and compliance is assured, the central variables of neoliberal institutionalism. In neoliberal anarchy, the worst-case scenario is a state's loss of gains due to the failure of cooperation; in realist anarchy, the worst-case scenario is a state's loss of its relative position and possibly a lack of survival in the international system.

The neorealist response to Keohane's theory of neoliberal institutionalism produced seminal works of its own. Kenneth Waltz laid the theoretical groundwork for the revival of realism in the 1970s through his article "Theory of International Relations" in 1975 and a full-length book in 1979 entitled *Theory of International Politics*. Greico's "Anarchy and the Limits of Cooperation: A Neorealist Critique of the Newest Liberal Institutionalism" in 1988 provided a comprehensive response to Keohane's theory from a realist perspective.[180]

The neorealist scholar Waltz, who engaged in an academic back-and-forth with Keohane over several years, described the international system as a "self-help" regime because, unlike the governments of nation-states, no single entity possessed a monopoly on violence.[181] As a consequence, actors must prepare to defend themselves because hostilities could break out at any time. In a self-help system, each unit spends a portion of its effort not in promoting its own welfare, but in providing the means to protect itself from others. While specialization in a system of divided international labor may benefit everyone, inequality in the expected distribution of the benefits may make states skeptical of cooperation. This is similar to Greico's conclusion: if one state may acquire a disproportionate benefit, even if both benefit, the uncertainty of each about the other's future intentions deters cooperation. Although the well-being of the system as a whole would benefit if states engaged in a more elaborate and specialized division of labor, doing so will make them ever more closely interdependent, and that makes them more vulnerable to external disruptions.[182] Consequently, even though they may benefit in absolute terms, states have an inherent tendency to avoid cooperation, especially if they are relatively small or weak in the global balance of power.

Self-help is a necessary principle of action in an anarchic order, and actors in an anarchic system must rely on the means they can generate in order to maintain their security. A self-help system is one of high risk, both of economic bankruptcy and of war, but organizational costs are low. Within an international order, risk can be mitigated through erecting agencies that possess effective authority and extend a system of rules. In hierarchic orders of supranational agencies, organizational costs are much higher, and the means of control becomes an object of struggle. States cannot entrust managerial powers to a central agency unless that agency is able to protect its client states. The more powerful the clients and the more powerful a threat each client appears to be, the stronger the central authority must be. The greater the power of the central authority, the stronger the incentive for states to struggle to control it.[183] In a society of states with little coherence, attempts at world government would founder on the inability of an emerging central authority to command resources to enforce compliance with the system. The prospect of world government is an invitation to world war. Units in an anarchic order act for their own sake and not for the sake of preserving an organization; consequently, states are free to leave one another alone. The system is peaceful in some sense: states are better able to aim for a minimal agreement that will permit their separate existence rather than a mutual agreement for unity's sake. "If might decides, then bloody struggles over right can more easily be avoided."[184] Wars among states cannot settle questions of authority or right; they can only determine the allocation of gains or losses for a time.

Keohane's response to Waltz and other neorealist critics is that, while an understanding of power dynamics in the international system is essential for any theory of international relations, realism does not provide a set of plausible and testable answers to questions about state behavior under specified conditions.[185] Realism helps determine the strength of the trap but does not provide assistance in escaping from the trap. In order to promote peaceful change, an international relations theory must not only determine the structure of world politics independent of individual decision makers, but must include variables that can be manipulated by human action. Since international institutions, norms, and patterns of cooperation can affect calculations of interest, and can also be affected by political decision making, they provide a natural focus for attention by scholars and policy makers.[186]

Keohane also points out a broader ambiguity in realist thinking. Assuming power to be tangible resources that can be used to induce other actors to do that which they could not otherwise do, balance of power theory is inconsistent with the realist assumption that states consistently seek to "maximize power."[187] States concerned with self-preservation do not seek to maximize power when they are not in danger. Rather, they recognize a trade-off between aggrandizement and self-preservation; they moderate their positions when their positions are secure, and they intensify them only when threatened. The realist assumption that states seek to maximize power does not explain why states weigh competing goals that are generated by their own internal social, political, and economic characteristics and only sometimes choose to maximize power.[188]

The theoretical division between relative gains and absolute gains was synthesized to some degree by Powell in an article that argues states are more sensitive to relative gains than to absolute gains when the possible use of force is implicated.[189] In this model, states are assumed to be trying to maximize their absolute gains, as a state's utility depends on the absolute levels of economic welfare the state gains. But the use of force may be at issue because states are trying to maximize their economic welfare within the constraints of an anarchic international system.[190] If the use of force is no longer at issue, however, a state's relative loss will not be turned against that state. When the cost of using force is sufficiently low that the use of force actually is at issue, cooperative outcomes that offer unequal absolute gains cannot be supported as part of the equilibrium even though the states' preferences are defined only over their absolute level of economic welfare. This inability to cooperate is in accord with the expectations of structural realism, though the assumption that states are maximizing their absolute gains is not in keeping with its usual formulations. If the use of force is not at issue because fighting is too costly, then the results are more in accord with neoliberal institutionalism.[191] This model suggests that cooperation, at least in the circumstance where the use of force is an option, may be especially difficult to achieve because cheating on the agreement would bring large relative gains. However, the model suggests that cooperation in other circumstances would be more promising than structural realists indicate.[192] Many aspects of neoliberal institutionalism and structural realism are actually special cases of the model.

The neorealist John Mearsheimer has criticized neoliberal institutionalism, believing that institutions have "minimal influence on state behavior, and thus hold little promise for promoting stability in the post–Cold War world."[193] To realists, institutions are basically a reflection of the distribution of power in the world, based on the self-interested calculations of the Great Powers, and consequently have no independent effect on state behavior. They are not important causes of peace and only matter at the margins. Because realist conceptions of cooperation are centered on absolute versus relative gains and on concern over cheating, cooperation is difficult to achieve and always difficult to sustain in a realist world.[194] The most powerful states create and shape institutions so that they can maintain or increase their share of world power, and institutions simply reflect the prevailing balance of power. Mearsheimer is likewise critical of Keohane's relative neglect of security issues in favor of environmental and economic issues, noting that neoliberal institutionalists overstate the propensity of states to cooperate because they exclude military issues from the analysis.[195] Cheating is a much greater obstacle to cooperation in the military realm than it is in the economic realm. In addition, other liberal counterarguments are not empirically sound. For instance, it has not been shown to be true that curing the cheating problem necessarily resolves the relative gains problem (on the theory that states would have nothing to fear from each other if they could trust one another not to cheat), and arguing that cheating is linked only to military affairs and not to economic ones creates a false separation between the two when military preparedness is so closely linked to economic strength. Scholars such as Keohane have been largely unable to point out instances in which cooperation would not have occurred but for institutions.

Jervis has found similar fault with Keohane's theory, but he walks back from Mearsheimer's "offensive realism," which posits that states are continually trying to maximize power. "Defensive realists" such as Jervis argue that states are often willing to settle for the status quo and are driven more by fear of other states than by desire to maximize their own power.[196] Conflict can be caused where goals among states are truly incompatible, or it can be caused where states take self-defeating measures or engage in an interactive process that generates conflict. Consequently, at least a subset of conflict may be unnecessary, because not all conflict represents real incompatibility in states' goals. However, neoliberal institutionalism is too optimistic when it posits that institutions increase cooperation, because their very existence in the first place occurs only when they would benefit from the institution in reaching their goals.[197] The difference between realists and neoliberals in this sense would be whether institutionalism is a cause or an effect of cooperation; neoliberals would choose the former, and realists the latter.

Keohane and Martin have responded to these criticisms. Institutionalist theory makes no distinction between security issues and political economy issues, because information transfer is the crux of the theory.[198] Realists contend that in an uncertain, anarchic world, states must assume the worst about others' intentions, but access to greater information could lead policy makers to policies that more nearly maximize utility. Institutionalist theory also has implications for distributional issues. States are often torn not only between whether to cooperate or not cooperate, but about the form that cooperation should take, and the existence of institutional frameworks may help states realize maximum potential gains from cooperation. Just as institutions can mitigate fears of cheating in the international system and allow cooperation to emerge, so too can they alleviate fears of unequal gains from cooperation.[199] Furthermore, realism simply cannot explain why states have invested so heavily in institutions such as GATT, NATO, and the European Union, if indeed these institutions are "marginal" to international politics.[200] Presumably, states would not support organizations that they believed to make no difference. Martin's research showed that states used the framework of the European Community to reduce fears of cheating by subverting sanctions on Argentina during the Falklands War.[201] European states generally refrained from sanctions busting even though they had the opportunity to profit from it.[202] Institutions do not just prevent or discourage cheating; they also allow for effective retaliation against cheaters and allow gains to be distributed among those who resist cheating.

A Neoliberal Institutional Theory of Integration

Integration in neoliberal institutional theory occurs when supranational institutions become autonomous enough from the self-interest of member states to encourage their own growth and begin to change state behavior toward further cooperation. Neoliberal institutionalism is the theory that cooperation among nation-states can be facilitated through the workings of international regimes or (more formally) supranational institutions, because both of these frameworks prevent defection, ensure access to greater information, and preserve reputations. Adopting the realist assumption that states have survival and self-interest as core motivations, states stand to benefit from joining international institutions. Institutions help states reach qualitatively more beneficial goals because they reduce transaction costs over the long term and allow for fruitful, cooperative relationships to develop among states. Even where an international regime is characterized by repetitive norms or practices rather than by a physical structure, such as the Law of the Sea regime, states can still benefit over the long term because they can set expectations based on the practices of other states. Finally, institutions can alter state behavior over time by making cooperation less costly, by punishing defectors through sanctions and other penalties, and by reinforcing norms of good behavior.

In the context of the integrative process among states, a neoliberal institutionalist perspective will look at the roles played by international institutions in encouraging integration through the reduction of transaction costs, penalties and compliance, and reputational benefits, such as maintaining creditworthiness. Like the other modern theories of integration, neoliberal institutionalism is not deterministic in its end goal, and it must account for reversal or disintegration. As state interests change, institutions may grow obsolete. Having developed through a long-running scholarly debate with revived neorealism, the realist–neoliberal institutionalist debate has helped both theories narrow points of contention and sharpen their essential elements. Because of increasing interdependence in the international sphere, Keohane indicated that realism understated the value and potential for cooperation among states. In an empirical study, a neoliberal institutionalist perspective will focus on the efforts of international institutions at encouraging interdependence among states, and at the response they receive from state officials and wider populations.

CONCLUSION

The three theories analyzed above share common theoretical origins in the project of Western European integration during the Cold War period, but each has attempted to refine its generally applicable principles so as to encompass regions outside Western Europe. With the constructivist revolt of the 1990s, security communities theory has faced a resurgence in scholarly literature that underpins the theory in sociological conceptions of individual and collective identity. The theory has been especially active in scholarship related to Southeast Asia and Southern Africa. Neofunctionalism has not fared as well in transitioning outside the Western European context, but certain elements emphasized in neofunctional scholarship, such as administrative and regulatory intertwinement and harmonization of policy across borders, continue to be useful in measuring the extent of regional integration in a particular case. Finally, neoliberal institutionalism is of more recent origin, developing in the late 1970s and early 1980s, but the theory benefited from the circumstances of the end of the Cold War, in which international institutions strengthened rather than withered with the end of the bipolar international system. Neoliberal institutionalism as a consequence may help explain why some international institutions have grown quite powerful and are able to shape state interests in varying degrees over time. The study of integration among Azerbaijan, Georgia, and Turkey that follows will continue to draw on these elements in analyzing the empirical evidence of integration.

NOTES

1. Joseph S. Nye, "Multinational Enterprises and Prospects for Regional and Global Political Integration," *The Annals of the American Academy of Political and Social Science* 403 (1973): 117–19.
2. Arend Lijphart, "Karl W. Deutsch and the New Paradigm in International Relations," in *From National Development to Global Community: Essays in Honor of Karl W.*

Deutsch, ed. Richard L. Merritt and Bruce M. Russett (London: Allen and Unwin, 1981), 233–34.

3. Karl W. Deutsch, *Political Community and the North Atlantic Area: International Organization in the Light of Historical Experience* (Princeton, NJ: Princeton University Press, 1957).

4. Donald Puchala, "Integration Theory and the Study of International Relations," in *From National Development to Global Community: Essays in Honor of Karl W. Deutsch*, ed. Richard L. Merritt and Bruce M. Russett (London: Allen and Unwin, 1981), 151.

5. Deutsch, *Political Community and the North Atlantic Area.*

6. In later works, Deutsch pointed to non-Western examples as well, such as partition between India and Pakistan and the Nigerian Civil War. Deutsch, *The Analysis of International Relations* (Englewood Cliffs, NJ: Prentice-Hall, 1968), 194–95.

7. Deutsch identifies only a single example where this has occurred in the past two hundred years: the disintegration of the German Confederation of 1815 that led to war between Prussia and Austria. However, he indicated that this likely was not a true security community. Deutsch, *Political Community and the North Atlantic Area*, 30–31.

8. Ibid., 31.

9. Note here a distinction from traditional realism, where states will oppose the rising strength of a core state in order to preserve a balance of power.

10. Deutsch, *Political Community and the North Atlantic Area*, 56.

11. Deutsch, *Analysis of International Relations*, 158–60.

12. Ibid., 195–96.

13. Ibid., 166–68, 198.

14. Helen S. Feldstein, "A Study of Transaction and Political Integration: Transnational Labour Flow within the European Economic Community," *Journal of Common Market Studies* 6, no. 1 (1967): 24–55.

15. Ibid., 48.

16. Ibid.

17. Puchala, "Integration Theory and the Study of International Relations," 151.

18. Frank Möller, *Thinking Peaceful Change: Baltic Security Policies and Security Community Building* (Syracuse, NY: Syracuse University Press, 2007), 13–14.

19. Ibid.

20. Ibid.

21. Emanuel Adler and Michael Barnett, "Security Communities in Theoretical Perspective," in *Security Communities*, ed. Emanuel Adler and Michael Barnett (New York: Cambridge University Press, 1998), 8–9.

22. Ibid.

23. Alex J. Bellamy, *Security Communities and Their Neighbors: Regional Fortresses or Global Integrators?* (New York: Palgrave Macmillan, 2004), 7.

24. Adler and Barnett, "Framework for the Study of Security Communities," in Adler and Barnett, *Security Communities*, 30.

25. Ibid., 43

26. Ibid., 50–53.

27. Ibid., 53–58.

28. Bellamy, *Security Communities and Their Neighbors*, 7–8.

29. Ibid., 10–11.

30. Ibid., 11.

31. Ibid., 178.

32. To that end, the broadened communications among members of a security community increase the amount of knowledge that a society has about another society. As Starr explains, "Such transparency means that each party has too much information about the other to create convincing enemy images for either elites or masses." The consequence is that societal support for hostilities against the democratic neighbor is reduced, and the society will be less likely to bear the costs of war. The opening of communications among members of a security community and the spillover consequences with nonmembers may reduce the prospect of war among them. Harvey

Starr, "Democracy and Integration: Why Democracies Don't Fight Each Other," *Journal of Peace Research* 34, no. 2 (2007): 157.

33. Möller, *Thinking Peaceful Change*, 36.

34. Dale C. Copeland, "The Constructivist Challenge to Structural Realism: A Review Essay," in *Constructivism and International Relations: Alexander Wendt and His Critics*, ed. Stefano Guzzini and Anna Leander (New York: Routledge, 2006), 1.

35. Alexander Wendt, *Social Theory of International Politics* (New York: Cambridge University Press, 1999).

36. This is not to say that Wendt's framework is not without shortcomings. He does not address issues of unequal communication or faulty information, i.e., on his spectrum of forms of culture and behavioral norms, a state may not know where another state is on the spectrum, which could hinder its own expectations of behavior. Copeland, "Constructivist Challenge," 18–19.

37. Möller, *Thinking Peaceful Change*.

38. Möller would not necessarily disagree with these propositions, but was more concerned about laying out the theoretical frameworks rather than the ideal end case. See James E. Goodby, Petrus Buwalda, and Dmitri Trenin, *A Strategy for Stable Peace: Toward a Euroatlantic Security Community* (Washington, DC: United States Institute of Peace, 2002).

39. Möller, *Thinking Peaceful Change*, 295.

40. Andrew Moravcsik, "'Is Something Rotten in the State of Denmark?': Constructivism and European Integration," *Journal of European Public Policy* 6, no. 4 (1999): 669–81.

41. Louis J. Cantori and Stephen L. Spiegel, "The Analysis of Regional International Politics: The Integration versus the Empirical Systems Approach," *International Organization* 27, no. 4 (1973): 476.

42. Amitav Acharya, *Constructing a Security Community in Southeast Asia: ASEAN and the Problem of Regional Order* (New York: Routledge, 2009), 2.

43. Acharya, *Constructing a Security Community*, 7.

44. Ibid., 31.

45. Ibid., 54ff.

46. In this, Acharya implicitly rejects Ole Waever's definition of a security community as simply a "no-war" community. See Ole Waever, "Insecurity, Security and Asecurity in the West European Non-War Community," in Adler and Barnett, *Security Communities*, 69–118.

47. Erik Martinez Kuhonta, "Walking a Tightrope: Democracy versus Sovereignty in ASEAN's Liberal Peace," *Pacific Review* 19, no. 3 (2006): 337–58.

48. Acharya, *Constructing a Security Community*, 39–42.

49. For instance, the demands of globalization (i.e., the widening gap between rich and poor, increased labor migration or refugee flows, etc.) may lead security communities to unravel. Increasing demands for democracy may also weaken the legitimacy of security communities composed of authoritarian states. Ibid.

50. Nikolas Busse, "Constructivism and Southeast Asian Security," *Pacific Review* 12, no. 1 (1999): 39–60.

51. Ibid., 56.

52. Benedikt Franke, "Africa's Evolving Security Architecture and the Concept of Multilayered Security Communities," *Cooperation and Conflict* 43, no. 3 (2008): 313–40.

53. The charter of the African Union, for instance, unlike its OAU predecessor, places limits on sovereignty. Franke links this to a changing sense of pan-African identity fueled by more cooperative attitudes after the end of the Cold War. Ibid., 322.

54. Ibid.

55. Laurie Nathan, "Domestic Instability and Security Communities," *European Journal of International Relations* 12, no. 2 (2006): 275–99.

56. This might be a slight overstatement. As Bellamy stated, membership in a collective security alliance is not necessarily coterminous with a security community. For instance, NATO had a strong relationship with Austria that could be considered a

security community even though Austria was for many years not a member state. On the other hand, Greece and Turkey are both members of NATO today but these two states are engaged in perpetual war preparation with one another. The point here is that even though SADC may have rough borders, it may be inconceivable for several of its members to go to war with one another (for instance, South Africa, Namibia, and Botswana). See Bellamy, *Security Communities and Their Neighbors*, 9–10.

57. Laurie Nathan, "The African Union and Regional Organisations in Africa: Communities of Insecurity," *African Security Review* 19, no. 2 (2010): 106.

58. Franke, "Africa's Evolving Security Architecture," 331.

59. Fotios Moustakis, *The Greek-Turkish Relationship and NATO* (Portland, OR: Frank Cass, 2003), 173–74.

60. Fotios Moustakis and Tracey German, *Securing Europe: Western Interventions towards a New Security Community* (New York: Tauris Academic Studies, 2009), 157.

61. This outcome would conflict with Bellamy's conclusions, summarized above. Presumably he would argue that as these peripheral regions become more tightly integrated with NATO and the EU, they begin to express their diplomatic relations in nonrealist terms and their cooperative attitude becomes contagious in a sense.

62. Deutsch, *Political Community and the North Atlantic Area*, 32–33.

63. Deutsch, *Analysis of International Relations*, 158–59.

64. This is Möller's definition. Möller, *Thinking Peaceful Change*, 13–14.

65. Arne Niemann and Philippe C. Schmitter, "Neofunctionalism," in *European Integration Theory*, ed. Antje Wiener and Thomas Diez (Oxford: Oxford University Press, 2003), 45.

66. Bastiaan van Apeldoorn, Henk Overbeek, and Magnus Ryner, "Theories of European Integration: A Critique," in *A Ruined Fortress? Neoliberal Hegemony and Transformation in Europe*, ed. Alan W. Cafruny and Magnus Ryner (New York: Rowman & Littlefield, 2003), 21.

67. Ibid.

68. Michael O'Neill, *The Politics of European Integration: A Reader* (New York: Routledge, 1996), 31.

69. Hans Morgenthau, introduction to *A Working Peace System*, by David Mitrany (Chicago: Quadrangle, 1968), 10–11.

70. Mitrany, *A Working Peace System*.

71. Ibid., 205.

72. Jennifer Sterling-Folker, "Competing Paradigms or Birds of a Feather? Constructivism and Neoliberal Institutionalism Compared," *International Studies Quarterly* 44, no. 1 (2000): 101.

73. Ibid.

74. David Long and Lucian M. Ashworth, "Working for Peace: The Functional Approach, Functionalism and Beyond," in *New Perspectives on International Functionalism*, ed. Lucian M. Ashworth and David Long (New York: St. Martin's, 1999), 1–26.

75. O'Neill, *Politics of European Integration*, 33.

76. Ibid.

77. Sterling-Folker, "Competing Paradigms or Birds of a Feather?," 102.

78. For the stealth metaphor, see O'Neill, *Politics of European Integration*, 35.

79. Constantin Stefanou and Helen Xanthaki, "The Principle of the Effective Protection of the Individual in EC Law and the Dialectic of European Integration Theory," *Northern Ireland Legal Quarterly* 50, no. 2 (1999): 215.

80. Michael Burgess, "Introduction: Federalism and Building the European Union," *Publius* 26, no. 4 (1996): 3.

81. For criticisms of Monnet's efforts, see ibid., 4.

82. Monnet is remembered for his lack of personal ambition and his comfort working with low-level technocratic staff in order to build loyalty from the bottom up; his solid personal instincts helped him to persuade political elites without being seen as a rival. George W. Ball, introduction to *Jean Monnet: The Path to European Unity*, ed. Douglas Brinkley and Clifford Hackett (New York: St. Martin's, 1991).

83. John Gillingham, "Jean Monnet and the European Coal and Steel Community: A Preliminary Appraisal," in Brinkley and Hackett, *Jean Monnet*, 129–62.
84. Ibid., 155.
85. Stephen A. Shuker, "The European Union: From Jean Monnet to the Euro," in *The European Union: From Jean Monnet to the Euro*, ed. Dean J. Kotlowski (Athens: Ohio University Press, 2000), 24–25.
86. Gillingham, "Jean Monnet and the European Coal and Steel Community," 157.
87. Shuker, "European Union," 27.
88. Ernst B. Haas, *The Uniting of Europe: Political, Social, and Economic Forces, 1950–1957* (Notre Dame, IN: University of Notre Dame Press, 1968), 33–37, 59.
89. Niemann and Schmitter, "Neofunctionalism," 46.
90. Philippe Schmitter, "Ernst B. Haas and the Legacy of Neofunctionalism," *Journal of European Public Policy* 12, no. 2 (2001): 255–72.
91. Ben Rosamond, "The Uniting of Europe and the Foundation of EU Studies: Revisiting the Neofunctionalism of Ernst B. Haas," in *The Disparity of European Integration: Revisiting Neofunctionalism in Honour of Ernst B. Haas*, ed. Tanja A. Börzel (Abingdon, UK: Routledge, 2006), 28–30.
92. Ibid.
93. Ibid.
94. Ibid.
95. As explained in Niemann and Schmitter, "Neofunctionalism," 48–49.
96. Philippe Schmitter and Ernst B. Haas, *Mexico and Latin American Economic Integration* (Berkeley: University of California Institute of International Studies, 1964).
97. Ibid., 32.
98. Schmitter and Haas did conclude, however, that spillover would be less likely in the Mexican context because the message of economic integration was so tightly controlled and centralized. In Western Europe, by contrast, the process of integration is broadly linked to long-range expectations of increasing welfare among large numbers of voluntary groups. Ibid., 33–34.
99. Leon N. Lindberg, *The Political Dynamics of European Economic Integration* (Stanford, CA: Stanford University Press, 1963), 5.
100. Ibid., 7–8.
101. Ernst B. Haas, "International Integration: The European and Universal Process," *International Organization* 15, no 3. (1961): 366–92.
102. Lindberg, *Political Dynamics of European Integration*, 293–94.
103. Sterling-Folker, "Competing Paradigms or Birds of a Feather?," 102.
104. Joseph S. Nye, "Patterns and Catalysts in Regional Integration," *International Organization* 19, no. 4 (1965): 870–84.
105. Ibid., 882.
106. Ernst B. Haas, "The Study of Regional Integration: Reflections on the Joy and Anguish of Pretheorizing," in *Regional Integration: Theory and Research*, ed. Leon N. Lindberg and Stuart A. Scheingold (Cambridge, MA: Harvard University Press, 1971): 618–19. By 1971, Haas was referring to neofunctionalism as a "pretheory" rather than a "theory."
107. Philippe C. Schmitter, "Three Neo-Functional Hypotheses about International Integration," *International Integration* 23, no. 1 (1969): 161–66.
108. Leon N. Lindberg, "Political Integration as a Multidimensional Phenomenon Requiring Multivariate Measurement," in Lindberg and Scheingold, *Regional Integration*, 649-731.
109. Ibid., 51.
110. Ibid., 55–56.
111. Niemann and Schmitter, "Neofunctionalism," 51–52.
112. Ibid., 49–50.
113. Ibid., 45–66.
114. Ibid., 52.

115. Catherine MacMillan, "The Application of Neofunctionalism to the Enlargement Process: The Case of Turkey," *Journal of Common Market Studies* 47, no. 4 (2009): 805.
116. Ibid., 52.
117. Ibid., 57–58.
118. O'Neill, *Politics of European Integration*, 43.
119. Schmitter, "Three Neo-Functional Hypotheses," 164.
120. Paul Stephenson, "Let's Get Physical: The European Commission and Cultivated Spillover in Completing the Single Market's Transport Infrastructure," *Journal of European Public Policy* 17, no. 7 (2010): 1039–57.
121. Ernst B. Haas, *The Obsolescence of Regional Integration Theory* (Berkeley: University of California Institute of International Studies, 1975), 86.
122. Ibid., 72.
123. Ibid., 17.
124. O'Neill, *Politics of European Integration*, 51.
125. Niemann and Schmitter, "Neofunctionalism," 63.
126. Alec Stone Sweet and Wayne Sandholtz, "European Integration and Supranational Governance," *Journal of European Public Policy* 4, no. 3 (1997): 297–317. At the same time, Sweet and Sandholtz reject Haas's belief that loyalties will eventually be transferred from the nation-state to the supranational entity. They reasoned that there was substantial room for supranational governance without an ultimate shift in identification. Ibid., 301.
127. Niemann and Schmitter, "Neofunctionalism," 61–63.
128. MacMillan, "Application of Neofunctionalism to the Enlargement Process," 789–809.
129. Cited in Niemann and Schmitter, "Neofunctionalism," 55.
130. Haas, "Study of Regional Integration," 24–25.
131. Ibid., 23–24.
132. Walter Mattli, "Ernst Haas's Evolving Thinking on Comparative Regional Integration: Of Virtues and Infelicities," in Börzel, *The Disparity of European Integration*, 118–19.
133. Stanley Hoffmann, "Obstinate or Obsolete? The Fate of the Nation-State and the Case of Western Europe," *Daedalus* 95, no. 3 (1966): 862–915.
134. Haas, "Study of Regional Integration," 24–25.
135. Amy Verdun, "Merging Neofunctionalism and Incrementalism," in *The Euro: European Integration Theory and Economic and Monetary Union*, ed. Amy Verdun (New York: Rowman & Littlefield, 2002), 11.
136. Verdun, "Merging Neofunctionalism and Incrementalism," 12–13.
137. Ibid., 27.
138. Ibid., 26.
139. Ernst B. Haas and Philippe Schmitter, "Economics and Differential Patterns of Political Integration: Projections about Unity in Latin America," *International Organization* 18, no. 4 (1964): 710.
140. Ibid.
141. Robert O. Keohane, "Neoliberal Institutionalism: A Perspective on World Politics," in *International Institutions and State Power: Essays in International Relations Theory* (Boulder, CO: Westview, 1989), 3.
142. David A. Baldwin, "Neoliberalism, Neorealism, and World Politics," in *Neorealism and Neoliberalism: The Contemporary Debate*, ed. David A. Baldwin (New York: Columbia University Press, 1993), 13.
143. Ibid.
144. Robert O. Keohane, *After Hegemony: Cooperation and Discord in the World Political Economy* (Princeton, NJ: Princeton University Press, 2005), x.
145. Keohane, *After Hegemony*, 245–46.
146. At least this was true at the time of Keohane's book. Arguably, the Kyoto regime is in some danger precisely because of American aloofness. Under the Obama admin-

istration, the United States has also reached a kind of détente with the International Criminal Court.

147. Ibid., 246.

148. Robert O. Keohane and Lisa L. Martin, "The Promise of Institutionalist Theory," *International Security* 20, no. 1 (1995): 40.

149. Keohane, *After Hegemony*, 244–45.

150. Andrew Moravcsik and Frank Schimmelfennig, "Liberal Intergovernmentalism," in Wiener and Diez, *European Integration Theory*, 72.

151. Quoted in Peter J. Katzenstein, Robert O. Keohane, and Stephen D. Krasner, "*International Organization* and the Study of World Politics," *International Organization* 52, no. 4 (1998): 660.

152. Ibid., 662.

153. Christopher Hemmer and Peter J. Katzenstein, "Why Is There No NATO in Asia? Collective Identity, Regionalism, and the Origins of Multilateralism," *International Organization* 56, no. 3 (2002): 576.

154. Ibid.

155. Ibid., 601.

156. Sterling-Folker, "Competing Paradigms or Birds of a Feather?," 103.

157. N. J. Rengger, *International Relations, Political Theory, and the Problem of Order: Beyond International Relations Theory?* (New York: Routledge, 2000), 128.

158. Ibid.

159. Robert O. Keohane and Joseph S. Nye, *Power and Interdependence: World Politics in Transition* (Boston: Little, Brown, 1977), 9.

160. Ibid., 35.

161. Joseph S. Nye and Robert O. Keohane, "Transnational Relations and World Politics: An Introduction," in *Transnational Relations and World Politics*, ed. Robert O. Keohane and Joseph S. Nye (Cambridge, MA: Harvard University Press, 1971), xvi–xix.

162. Ibid., xix–xxii.

163. Ibid., 374–75.

164. Keohane, *International Institutions and State Power*.

165. Ibid., 11.

166. Ibid., 117.

167. Ibid., 120–22.

168. Ibid., 123–24.

169. Ibid., 112–13.

170. Lisa L. Martin, "Interests, Power, and Multilateralism," *International Organization* 46, no. 4 (1992): 765–92.

171. Ibid., 782–83.

172. Robert Axelrod and Robert O. Keohane, "Achieving Cooperation under Anarchy: Strategies and Institutions," in Baldwin, *Neorealism and Neoliberalism*, 109–10.

173. Ibid., 110.

174. Ibid., 111.

175. Joseph M. Greico, "Anarchy and the Limits of Cooperation: A Realist Critique of the Newest Liberal Institutionalism," *International Organization* 42, no. 3 (1988): 485–507.

176. Ibid., 487.

177. Baldwin, "Neoliberalism, Neorealism, and World Politics," 4–8.

178. Ibid., 6.

179. Greico, "Anarchy and the Limits of Cooperation," 497–98.

180. Baldwin, "Neoliberalism, Neorealism, and World Politics," 13.

181. Kenneth N. Waltz, "Anarchic Orders and Balances of Power," in *Neorealism and Its Critics*, ed. Robert O. Keohane (New York: Columbia University Press, 1986), 98, 100.

182. Ibid., 102–3.

183. Ibid., 110.

184. Ibid.
185. Robert O. Keohane, "Theory of World Politics: Structural Realism and Beyond," in Keohane, *Neorealism and Its Critics*, 159.
186. Ibid., 199.
187. Keohane, *International Institutions and State Power*, 47.
188. Ibid.
189. Robert Powell, "Absolute and Relative Gains in International Relations Theory," *American Political Science Review* 85, no. 4 (1991): 1303–20.
190. Ibid.
191. Ibid.
192. Ibid., 1305, 1316.
193. John J. Mearsheimer, "The False Promise of International Institutions," *International Security* 19, no. 3 (1995): 7.
194. Ibid., 12.
195. Ibid., 16.
196. Robert Jervis, "Realism, Neoliberalism, and Cooperation: Understanding the Debate," *International Security* 24, no. 1 (1999): 42–63.
197. Ibid., 54.
198. Keohane and Martin, "Promise of Institutionalist Theory," 39–51.
199. Ibid., 45.
200. Ibid., 47.
201. Ibid., 49.
202. The Falklands example had been dismissed by Mearsheimer in "The False Promise of International Institutions," 25.

TWO

Evaluating Integration among Azerbaijan, Georgia, and Turkey

The three integrative theories presented in the first chapter provide a framework for assessing relations among Georgia, Azerbaijan, and Turkey, three countries that have become increasingly entangled in certain areas since the fall of the Soviet Union. Representing highly distinct populations and varying greatly in terms of economic stability and political and legal capacity, an organic process of integration is nonetheless occurring, especially in the energy and transport sectors. This is no accident: energy and transit are highly regulated fields, and the intertwinement of the regulatory and administrative capabilities of all three countries reduces transaction costs and improves profits. The spillover process first delineated in traditional neofunctionalist theory, however, has not occurred. While regulatory reform undoubtedly improved transparency and legal capacity, it does not appear to have placed significant integrative pressure on other sectors, such as security or the economy. At the same time, there is mixed evidence of increasing communications among the Turkish, Azerbaijani, and Georgian populations to an extent greater than their communications with other countries in the region and with the world.[1] The involvement of transnational extra-regional actors (TERAs), such as the European Union and the World Bank, as facilitators of integration provides a more promising thesis. Nonetheless, all three theories of integration examined above provide a helpful framework for understanding relations among Azerbaijan, Georgia, and Turkey. The following chapter will evaluate integration among Azerbaijan, Georgia, and Turkey, during the period from the mid-1990s to 2008, in the spheres of security, politics, society and culture, economics and trade, energy, and transport. It will demonstrate that during the above time period, integration only occurred in the energy and transport sectors, before end-ing with an introduction to the role of TERAs in these processes, which will be more fully discussed in subsequent chapters.

SECURITY

The limits of integration among Azerbaijan, Georgia, and Turkey are evident in the security sector. During the 2008 war between Russia and Georgia, for instance, Turkey did not have the capacity to serve as a counterweight to Russia and, while quietly supportive, Azerbaijan very much remained on the sidelines. All three countries have maintained multiple security alliances and preferences, and all three have ambiguous political and economic relations with Russia, their largest trading partner during the time period examined. Similarly, the isolation of Armenia appears to be a higher priority for Azerbaijan and Turkey than it does for Georgia, while Azerbaijan's ambiguous relationship with Iran complicates the security alliance among the three countries even further.[2] Indeed, using Deutsch's definition of a security community, it is not clear at all that among Georgia, Azerbaijan, and Turkey an exclusive "we-feeling" and sense of common identity have developed. Instead, each country's security position has varied depending on the political leadership at the time.

Among the South Caucasus countries, Azerbaijan seemed the most likely to sustain long-term economic growth at the time of independence. In part this was because Azerbaijan had the potential to use its oil revenues to sustain a military challenge, although war broke out in Nagorno-Karabakh before the Azerbaijani oil sector was developed enough to fuel a military victory. Nonetheless, oil remains vital to Azerbaijani security and, as the progress of war with Armenia showed, Azerbaijan possesses a powerful weapon in cutting off gas supplies and blockading food and consumer goods along the Baku-Yerevan railway line, with devastating results for Armenia. Georgia, on the other hand, did not have Azerbaijan's resource potential, and possessed ambiguous relations with Armenia since independence. Georgia does not maintain an economic blockade on Armenia and in many ways acts as a quiet thoroughfare between Russia and Armenia, despite its important energy relationship with Azerbaijan. Georgia may in fact view Armenia as a strategic mediator of sorts in its own conflict with Russia. In 2008, Armenia maintained "constructive neutrality" toward Georgia during the Russia-Georgia War, but suffered after the war when Georgia and Russia failed to reopen border crossings to permit the export of Armenian produce.[3] Armenian-Georgian relations have been characterized as grudging détente, but in fact there is little that pits Tbilisi and Yerevan directly against one another on security matters.[4]

Russia's relationship with the countries of the South Caucasus is complex, ambiguous, and inconsistent. While Russia has tried to establish a security system with the independent states of the South Caucasus, it has only succeeded to a significant degree in Armenia, the only former Soviet republic to have a combined air-defense system with Russia.[5] Surrounded by hostile neighbors, Armenia does not need to be pressured to convince Russian troops to remain on its soil. Georgia was initially reluctant to permit a base for Russian troops, given Russian support for separatists in Abkhazia and South Ossetia, but it later signed an agreement allowing Russia to police the border with Turkey and hosted Russian military facilities until the lead-up to the 2008 war.[6] For the first decade after Azerbaijan's independence, Russia pressured the country to allow Russian military bases on its soil, to closely monitor borders with Turkey and Iran, and to lease the early-warning station at Gabala in northeast Azerbaijan. Azerbaijan, however, has consistently rejected the stationing of Russian troops on its soil due to perceived Russian support of Armenia in the Nagorno-Karabakh dispute. Azerbaijan has attempted to use the early-warning radar station at Gabala as a bargaining chip to obtain Russian support on Karabakh, but without much success.[7]

Russia's own security concerns in the South Caucasus are intimately tied to cross-border ethnic and clan linkages in the North Caucasus region, which was incompletely "colonized" by Moscow during Soviet times. Azerbaijan's proximity to Chechnya, a restive autonomous region in the North Caucasus, and its border with Dagestan, another potentially explosive region, threaten Russia's careful security policy. Similarly, the war in South Ossetia in Georgia created the conditions leading to arms trafficking and organized criminal activity in North Ossetia and Ingushetia in the Russian Federation. The Georgian war in Abkhazia generated unrest in the ethnically related Russian autonomous republic of Adyghe.[8] As a consequence, Russia vigilantly monitors its southern border with Georgia and Azerbaijan. While Dagestan may be less likely to suffer a Chechen-style ethnic conflict, the Russian pipelines from Baku to Novorossiysk and from Baku to Tikhoretsk pass through the Dagestani capital of Makhachkala. The strategic importance of these routes is even greater than the Chechen routes: while a loop around Grozny is feasible, transit through Dagestan is unavoidable as it is the only border Russia possesses with Azerbaijan.[9] A Chechen scenario in Dagestan could have dire consequences for the viability of Russian routes for the export of oil from Azerbaijan.

To the extent that a Russia-Armenia-Iran axis exists in the South Caucasus region (exclusive and exhaustive of a Georgia-Azerbaijan-Turkey axis), it is also the product of ambiguous relations between Iran and Azerbaijan. While Iran took a more ideological approach toward Azerbaijan shortly after independence and emphasized an affinity for the Muslim Azeri population during the Nagorno-Karabakh war, Iran has been increasingly vigilant of destabilization and ethnic unrest among Azeri populations in northern Iran.[10] The more Baku expressed a desire for ties with Iran's Azeri populations, the more Iran expanded its cooperation with Armenia.[11] When President Elchibey of Azerbaijan, for instance, called for increasing ties with "South Azerbaijan" (northern Iran), Iran established diplomatic relations with Armenia and improved natural gas and transport links with Armenia at the height of the Nagorno-Karabakh war. By transporting Russian oil through Iranian territory, Iran broke the Azerbaijani blockade and Turkish embargo and sustained Armenia's war effort.[12]

The expansion of Russian influence in the South Caucasus has fueled in large part the potential for an Azerbaijan-Georgia-Turkey axis. Turkey in particular is concerned that Russia may use the conflict in Chechnya as a pretext to reestablish greater influence, if not control, over the Southern Caucasus. This would place Turkey in a security dilemma of sorts, exacerbating the country's security fears by endangering an important land corridor between Turkey and Central Asia and threatening plans to build a trans-Caucasian network of pipelines. Turkey is not an equal counterweight to Russia. Although Turkey would not be able to single-handedly prevent Russian encroachment on the Caucasus, Ankara would object to an expansion of Russian influence and would fear damage to Turkey's strong economic relations with Russia.[13]

Turkey's close relations with Azerbaijan and Georgia have resulted not only from a more nuanced foreign policy, but also from the resolution of significant internal crises that plagued Turkey during the 1990s. Turkey's improved relations with its neighbors and the European Union foreshadowed a rise of influence for the country in the Caucasus. In 1997, Turkey and Georgia signed a defense cooperation agreement that led to, for instance, Georgian troops serving under Turkish command in the United Nations force in Kosovo.[14] Turkish support following a renewed cooperation agreement in 2000 included $2.8 million for the construction of military training centers in Kodori and Gori and a shooting range outside Tbilisi, as well as training of Georgian pilots in Turkish airspace.[15] Turkey spent $1.5 million modernizing the Marneuli military air base outside Tbilisi, opened in 2001, which had the symbolic value of being the first modern base to be established in Georgia without Russian assistance.[16] As a consequence of this period of emerging stability in the Caucasus, by the year 2000 Turkey was seeking a major expansion of influence in the region by pursuing a détente with Russia. Turkey has sponsored several regional strategic initiatives in which Russia was also included, such as the Black Sea Economic Cooperation (BSEC) program and a possible Caucasus Stability Pact, although the latter has not progressed beyond a notional vision.[17] The Caucasus Stability Pact, proposed by then-Turkish president Süleyman Demirel, was revived by Prime Minister Recep Tayyip Erdoğan in 2008 after the Russia-Georgia war as the Caucasus Stability and Cooperation Platform, but that was also unrealized.[18] Similarly, although many Turkish politicians individually possess a sense of kinship with the Central Asian Turkic republics, "pan-Turkism" has not been official government policy to avoid friction with Russia and China.[19]

Azerbaijan and Turkey have grown closer on security issues as the Nagorno-Karabakh issue continues to fester and dominate Azerbaijani politics and as Azerbaijan attempts to construct a professional military and security force with its newfound extractive resource wealth. For Azerbaijan, one of the most serious negative consequences of Soviet rule was the lack of military experience, as highlighted by Azerbaijan's serious military defeats in the Nagorno-Karabakh war. Moscow distrusted the Azerbaijanis more than the Armenians, who were provided better training, equipment, and field commanders. President Mutalibov was sluggish in organizing a national army, and a major Armenian offensive against Shusha and Lachin led to devastating defeats for the Azerbaijani army, including the offensive at Khojaly in which scores of people were killed in a final wave of ethnic cleansing from the oblast. The Mutalibov regime could not survive the grisly images produced by the massacre.[20]

Neither Georgia nor Azerbaijan has the luxury of completely eliminating their economic dependence on Russia in some areas. Western priorities toward Georgia do not always align with Georgian perceptions of Western priorities. Western security policies toward Georgia have come to focus on three main objectives: securing energy supplies, achieving regional stability and integration through the resolution of local conflicts, and strengthening Georgian sovereignty and governance transformation.[21] While Georgian leaders have heightened expectations about joining the Euro-Atlantic community, NATO priorities are for Georgia to engage in legal and economic reform and push for resolution of its internal conflicts.[22] From the standpoint of many NATO members, Georgia will remain dependent on Russia in many areas, and must develop durable solutions to internal and regional problems in cooperation with Russia. The same is likely true of Azerbaijan, although the country has broader range of motion because of its natural resource wealth.

Evaluation

To evaluate the extent to which integration has occurred among Azerbaijan, Georgia, and Turkey, it is necessary to emphasize what security integration is not: a realist-style alliance erected in response to an external threat. While the presence of a stronger, latently hostile neighbor such as Russia may nonetheless spur the creation of an integrated security community, supranational agency, or neofunctionally intertwined cluster, an external hostile neighbor is not a necessary or sufficient condition for security integration. Using Deutsch's definition of a pluralistic security community, namely (1) shared values, (2) the capacity of the participating political units to respond to one another's crises, and (3) mutual predictability of behavior, it is clear that the South Caucasus states fall far short. The unilateral behavior of Georgia in particular in triggering the 2008 Russia-Georgia war, and Russia's vigorous and disproportionate response, was far from the predictability of behavior that Deutsch envisioned. Likewise, Turkey's tepid response to Russia, by now its largest trade partner and energy supplier, undermined any realistic security arrangement among the three countries.[23] With important implications for neofunctionalist theory and neoliberal institutionalism, both the multinational oil companies that possessed a vested interest in regional peace and had worked so closely at developing an intertwined energy and transport sector as well as the supranational agencies that had laid so much of the groundwork for economic integration stood helplessly aside when war broke out in 2008. If security integration had occurred among the three countries to any extent, the Russia-Georgia war proved that it was superficial, even trivial.

Turkish-backed aspirations at creating a security alliance of sorts with Georgia and Azerbaijan, expressed through the proposed Caucasus Stability Pact and other theoretical initiatives, lay in ruins after five days of hot hostilities between Russia and Georgia. As Torbakov explains, the war wrecked Turkish policy toward the Caucasus of encouraging large-scale cooperation with Russia while engaging in subtle competition for control of resources and the transport of resources through Russia's neighbors.[24] Turkey had signed a limited defense-cooperation pact with Georgia and Azerbaijan, and had given Georgia $1.8 million in military aid in 2006 and provided some military training for Georgian officers. As of 2005, Turkish military aid to Georgia amounted to about $70 million, and was seen as an attempt to integrate Georgia into the European and Atlantic security networks.[25] This aid was used in restructuring the military apparatus, training personnel in technical infrastructure and equipment that comply with NATO standards, and reforming military academies. It has also been used to improve border security.[26] Likewise, Turkey's military assistance to Azerbaijan has exceeded $200 million as of 2010, including long-term military training and, most recently, a helicopter reconnaissance project.[27] By 2008, however, these efforts at security integration were simply too marginal to change the status quo in the region. Indeed, security cooperation among the three countries may be increasingly undermined by Turkey's growing dependence on Russia for its energy needs and by Russia's increased leverage over Turkey due to its favorable balance of trade.

In addition to the obvious absence of a security community among Azerbaijan, Georgia, and Turkey, neither the neofunctionalist perspective nor the neoliberal institutionalist perspective evidence any significant movement on integration among the three countries. This is in sharp contrast to Armenia, which shares a common air-defense system with the Russian Federation. Azerbaijan and Georgia's security links with Turkey are less extensive than Azerbaijan's with Russia, due to the Gabala radar station, or Turkey's with Western European countries through NATO. Georgia has NATO membership aspirations, but its security realities are in fact dictated by the Russian troop presence within its borders, in Abkhazia and South Ossetia. Russia may also have been involved in the coups d'état that brought Eduard Shevardnadze and Heydar Aliyev, the dominant postindependence national leaders in Georgia and Azerbaijan, respectively, to power, even though both leaders developed ambiguous relations toward Russia.[28] Even on the issue of Nagorno-Karabakh, arguably the neutralization of Russia's pro-Armenian position may have had a greater impact on the establishment of and preservation of a cease-fire than Turkey's more confrontational policies. Georgia, Azerbaijan, Turkey, and Russia are all members of the Organization for Security and Cooperation in Europe (OSCE), but otherwise none of the countries share common membership in a security organization such as NATO or regional organizations governing the use of chemical, nuclear, or biological weapons.[29] Security cooperation is hampered by the lack of a common security strategy, or even compatible security goals, among Azerbaijan, Georgia, and Turkey. Azerbaijan's security objectives are not necessarily as pro-European as Georgia's or Turkey's, and it does not aspire to NATO and EU membership to the same degree. Turkey does not relish the opportunity to become the security guarantor of Georgia and Azerbaijan, two countries in somewhat unstable security situations. On the issue of security integration, Azerbaijan, Georgia, and Turkey have not fulfilled even the most basic prerequisites of cooperation, and it is not clear that the three countries are even moving in the same direction.

Turkey, Azerbaijan, and Georgia have a considerable way to go before security integration is empirically noticeable. Although Turkey has been a sponsor of military aid to both Azerbaijan and Georgia consistently since the fall of the Soviet Union, it was unable to play more than a marginal role at the most critical point, that is, during the Russia-Georgia war of 2008. Security goals among the three countries overlap but do not completely coincide, as is evidenced by Georgia's ambiguous relations with Armenia. While hostilities among Turkey, Georgia, and Azerbaijan seem to be unlikely in the current geopolitical situation, the unmistakable reality is that Georgia was very much on its own when hostilities flared with Russia. Beyond a superficial layer of military aid and aspirational security promises among the three countries, little actual security integration has occurred. Certainly, the "we-feeling," the sense of mutual trust among the three countries' populations described by Deutsch in the prior chapter, has not developed. Even a "loosely coupled security community," described by Adler and Barnett as a transnational region comprised of sovereign states whose people maintain dependable expectations of peaceful change, fails to describe the security situation in the region. At a fundamental level, as the 2008 hostilities showed, the three countries do not possess a common perspective toward belligerent outsiders that is contemplated by even the loosest forms of a security community.

POLITICAL

This section will focus on the integration of political structures in Azerbaijan, Georgia, and Turkey, including harmonization of foreign relations, geopolitical orientations, and affinities with supranational political organizations. The three countries undoubtedly share some overlap on each of these variables, but the results are mixed and it is difficult to discern a single trend. As with security integration, the relationship among Azerbaijan, Georgia, and Turkey reveals little evidence of political integration, though with less clear-cut results. Each country maintains divergent foreign policy positions not only with Russia but also with Armenia and Iran.[30] Georgia alone has an open border with Armenia after the Nagorno-Karabakh war, and Azerbaijan possesses a kindred spirit but somewhat ambiguous relationship with Iran that differs from the more Western-oriented policies of Georgia and Turkey.[31] From a neoliberal institutionalist perspective, it is also important to note that the three countries are not united in geopolitical orientation. While they each possess common memberships (or aspirant memberships) in supranational communities, this is not universally true, and indeed the most salient of these communities, such as the European Union, NATO, or the Commonwealth of Independent States (CIS), excludes one or another of the three countries studied here. At the same time, these three countries are at different points in democratic development. Certainly, administrative and bureaucratic reform and anticorruption efforts have helped integrate the transport, telecommunications, and energy sectors and have lowered trade barriers. However, this increased regulatory transparency has not fundamentally altered the political orientation of the Southern Caucasus states or their foreign policies.

Politically, the three countries have an ambiguous relationship with Russia and with the West, and this relationship heavily influences their relations with one another. A common aversion to Russia may unite the three countries at the broadest level, but the similarity is superficial, as all three countries have quite distinctive relations with Russia. Georgia's recent history with Russia has been especially combative, and the Russian relationship with Georgia is seen as among the worst of any former Soviet republic, leading to hostilities in 2008. Azerbaijan has pursued a middle path between Russia and the European Union, strategically playing them off against each other and using its strong energy resources as leverage. Turkey, by contrast, has seen its relationship with Russia improve over the past decade as trade rapidly expands between them. Undoubtedly, political and legal reforms in all three countries have helped to curtail illicit border flows and corruption, which facilitated the integrative process, particularly with regard to energy and transport. However, whether a shared political identity is developing among the three countries is a more difficult question, given their divergent interests and relationships in the region.

Undoubtedly, Turkey and Azerbaijan have grown politically closer for sentimental reasons, including—at least until very recently—a shared commitment to secular Islam and Turkic culture, as well as realpolitik reasons—in particular, the conflict over Nagorno-Karabakh. Writing shortly after the Nagorno-Karabakh conflict, Cornell notes that Turkey was Azerbaijan's most significant defender, explaining and promoting the Azeri view of the conflict and jointly with Azerbaijan placed an embargo on Armenia.[32] Turkey also refused to normalize relations with Armenia as long as Armenia occupies territories in Azerbaijan.[33] This cooperation, however, has had its limitations. "However, despite its support for the Azeri cause, Turkey has largely stopped short of furnishing Azerbaijan with weapons or affluent financial aid which would enable it to buy any; nor has Turkey even threatened to intervene militarily on Azerbaijan's side."[34] Despite its support for Azerbaijan's position in the conflict, in other words, Turkey was not an entirely free agent.

Western powers perceived Turkey as a pivotal state for increasing Western influence in Azerbaijan and the Central Asian republics, due to its ethnic kinship with the Muslim peoples of the former Soviet Union and its prosecular model of democracy for the newly independent states. As a consequence, the first Bush administration in particular favored expansion of Turkish influence, in part as an antidote to a hostile Iran.[35] Turkey has an enviable geographic position, as a natural link between Europe and the South Caucasus and Central Asia, through the natural gateway of Azerbaijan. Turkey hailed the independence of the Central Asian republics, envisioning a Turkic common market of sixty million people, stretching from the Aegean Sea to China. Azerbaijan was a critical link to this goal, as it lies within geographical proximity to Turkey, shares an ethnic and linguistic kinship, and possesses enviable natural resources.[36] Turkey was the first country to recognize independent Azerbaijan and initiated high-level government contacts and economic relations with Baku, including an emphasis on cultural links through education and television broadcast programming.

However, Turkish aspirations at leadership of the Turkic world encountered real limitations in Turkey's own political willpower and influence. During the late 1990s, Turkey went through internal crises with Kurdish separatists and economic recession at a time when Russia was attempting to expand its influence over the same areas championed by

Turkey. Turkey is cut off from the rest of the Turkic world except for a short border with Nakhchivan, the Azerbaijani enclave, and was unprepared for the independence of the five Turkic republics. Turkey was indifferent to Soviet Turks for decades, and developed little expertise or research institutes on Soviet or Turkic affairs, and consequently, most Turkish aid to the new republics has been in the form of cultural ties. Despite a strong industrial base and Turkey's traditional status as a bulwark against Iran and Islamism and a model Muslim democracy, Turkey has not provided the level of investment and aid that the Turkic republics have sought.[37] In particular, when Turkish support and influence faced its most critical challenge—keeping the Turkish-backed Elcibey in power in Azerbaijan in the face of a coup—Turkey faded. The strong, pan-Turkic vision expounded by Turkish leadership eventually wore down as geopolitical realities dampened the early enthusiasm.

Immediately after independence, Azerbaijan indicated a determined desire to break with old Communist structures and adopt democratic institutions. In contrast with the more conservative Central Asian states, prospects for democratization in Azerbaijan seemed especially promising, as Azeri nationalism was well developed in its conceptual, political, and psychological aspects.[38] The events of summer 1993 indicated the resiliency of an authoritarian trend, as well as a turn away from newly established democratic ideals to long-existing clan and local ties. Aliyev's rise, however, was not a return to Communism, and he was welcomed across the political spectrum by many major political groups in Azerbaijan.[39] Azerbaijani foreign policy has grown more sophisticated to the point that the country now tries to balance Western and Russian interests to its own benefit. The country expressed to Russia that it had no ambition to join NATO, unlike Georgia, and made overtures to Russian President Dmitri Medvedev shortly after he took office, while US Vice President Richard Cheney faced an arguably cooler reception later in 2008.[40]

Almost immediately after independence, Azerbaijan faced an existential threat that determined the course of its development over the next two decades. Armenian military victories weakened the government of Azerbaijan. The fall of Shusha, the last town under control of Azerbaijan in Nagorno-Karabakh, led to the resignation of President Mutalibov and the election of Elchibey in June 1992. Espousing pan-Turkish sentiment, Elchibey refused to join the CIS and called for the removal of all Russian troops; he also signed the first accord for the establishment of a Baku-Ceyhan pipeline, to the irritation of Russian interests.[41] Elchibey was forced out of office in a Russian-backed coup, and replaced by Heydar Aliyev, a former member of the Soviet politburo.

Relations between Turkey and independent Azerbaijan predated relations with and political and economic interest in Georgia. In spring 1991, Turkish President Turgut Özal was the first Turkish president to visit Baku, partially in an effort to make amends over Özal's earlier comments that highlighted the differences between Sunni Turkey and Shi'a Azerbaijan. Azerbaijani President Ayaz Mutalibov visited Ankara in January 1992 to sign a treaty of friendship and cooperation. In May 1992, Turkish Prime Minister Demirel opened a bridge to connect Turkey to Nakhchivan and spoke to the Azerbaijani parliament in Baku.[42] Although Azerbaijani leaders have looked to Turkey and the traditional secular Turkish model of development, the closeness of the two countries reached a zenith under President Elchibey.

Turkey and Russia were headed for a collision course over Azerbaijan. After Elchibey was overthrown by a Russian-backed coup and Aliyev took power, relations between Turkey and Azerbaijan were initially cool, particularly as Azerbaijan joined the CIS in return for Russian support on Nagorno-Karabakh, and Aliyev withstood a coup attempt that was linked to some Turkish officials. Turkish citizens were obligated to obtain visas before entering Azerbaijan, and Turks without visas were rounded up and deported.[43] In addition, the Nagorno-Karabakh dispute caused some strain with Turkey. While Turkey sent 1,600 Turkish military experts to Azerbaijan, officials preferred an even-handed approach to the dispute so as not to inflame the Armenian diaspora lobby in the United States or spur a direct confrontation with Russia.[44] Aliyev dismissed some of the Turkish military experts and canceled an agreement committing Azerbaijan to the Baku-Tbilisi-Ceyhan pipeline. However, a ceasefire in Nagorno-Karabakh and Aliyev's refusal to station Russian troops in Azerbaijan improved relations with Turkey.[45] In general, the difficulties between Azerbaijan and Turkey were ephemeral episodes, however, and Turkey remained Azerbaijan's foremost ally in the region.

Elchibey's overthrow was undoubtedly a setback for integration among Azerbaijan, Georgia, and Turkey. Turkish support for Elchibey and for the opposition in Azerbaijan alienated the new rulers and encouraged them to look to Russia for support, and the Turkish failure to keep Elchibey in power signaled weakness to the other Turkic republics in Central Asia and undermined Turkey's image as counterweight to Russian influence.[46] The Western-backed Turkish model of promoting secularism and market-oriented economics had some success, especially for cultural support and occasional summits among the Turkic republics. In general, however, Turkey's economic weakness during the 1990s hampered the country's efforts to be a trendsetter, and grandiose Turkish ideas such as a Turkic common market proved to be more utopian than realistic.[47]

Though relations between Azerbaijan and Turkey are not as idealized as in Azerbaijan's early independence period because Azerbaijan sought a closer relationship with the United States and direct Turkish influence waned under President Aliyev, the stable relationship between the two countries has grown much deeper and more sophisticated. Turkey has increasingly become a consumer of Azerbaijani gas and Caspian hydrocarbon exports. In a conflict-ridden region, more importantly, the relationship between Turkey and Azerbaijan in particular was able to attract Western investors, who needed guarantees from the host governments that they would support freedom of oil transit and pipeline security. These commitments were in place in the intergovernmental agreement underpinning the Baku-Tbilisi-Ceyhan (BTC) pipeline even before extraction began in the Caspian. Cooperation between Azerbaijan and Turkey has also kept the issue of Russian bypass pipelines higher on the American agenda.[48]

The geography of oil in Azerbaijan contributes to political disintegration. Unlike Kazakhstan, where major oil deposits lay hundreds of miles away from the country's political and financial centers, Azerbaijan's richest oil fields are concentrated in and around the Absheron Peninsula, where Baku is situated. In an economy with a single valuable resource such as Azerbaijan's, the presence of oil in the political center prevents resource-based bargaining between Baku and the provinces, and consequently more peripheral regions of the country have suffered greatly without any voice in the ruling elite, including lack of basic government services.[49] Energy resources in Azerbaijan thus reinforce central authority, contributing to the zero-sum nature of domestic politics and the bitterness of political opposition. As a consequence, in the absence of resource-driven incentives to cooperate and share resources, political conflict focuses more on seizing and maintaining power. The consequence of this process for political integration in the region is that it creates resentment from the periphery and accentuates already-existing fault lines in Azerbaijan on the basis of geography, minority status, and distribution of wealth. Overshadowing Azerbaijani politics, however, to a degree far greater than energy or center-periphery relations, is the Nagorno-Karabakh conflict.[50]

Despite the prospects for political integration in the South Caucasus, Georgia in particular has faced a political volatility not shared by Azerbaijan or Turkey. The new source of development and revenue for Georgia in the form of oil and gas transit has not boosted prospects for democratization. Although Georgia briefly had a parliamentary system after independence, a strong presidential system has dominated politics in that country, with opposition forces generally weak or nonexistent since that time. Georgia was among the wealthiest republics in the Soviet Union in terms of per-capita consumption and real income, but suffered what was almost certainly the greatest economic decline of any former Soviet republic after independence.[51]

Russian influence has, at times, drawn Azerbaijan and Georgia apart. Azerbaijani-Georgian relations have thawed since Putin came to power in Russia, and he developed relatively close relations with Aliyev, who, like Putin, was a former KGB officer, in striking contrast to the cool relations between Putin and Shevardnadze. In 2000, Russia placed stricter visa requirements on Georgia while waiving them for Azerbaijan. In 2003, Russia's warm congratulations to Ilham Aliyev, son and successor to the previous president, who was elected in disputed elections, was in sharp contrast to Russia's cold ten-day silence when Saakashvili was elected in Georgia in an undisputed election in January 2004.[52]

Similarly, relations between Turkey and Georgia were initially cautious as political turmoil in Georgia—stabilized only when Shevardnadze took the chairmanship of the State Council in March 1992—delayed Turkish recognition until May 1992. Officials in Tbilisi suspected Turkey of supporting the possible ambitions of Muslim-populated Ajaria in southern Georgia.[53] At the time, Georgia's role as transit country for Caspian oil and gas was not yet recognized, but Shevardnadze encouraged regional cooperation by inviting Iran, Russia, and Turkey into a cooperative pact.[54] Turkey maintained a strong interest in the Abkhazia dispute and dispatched officers to participate in the United Nations Observer Mission in Georgia (UNOMIG), which monitored the cease-fire line.[55] Turkish diplomats emphasize the importance of Georgia's territorial integrity, particularly with the investment in pipeline transit that Turkey has made.[56] However, Turkey's relationship with both Azerbaijan and Georgia has stagnated in recent years. The Rose Revolution in Georgia raised the prospect of Georgia becoming a competitor with Turkey for EU membership, and the war in Iraq soured relations between the United States and Turkey. In Azerbaijan, the succession crisis at the death of President Heydar Aliyev caused unease in both Turkey and Russia, although both countries supported Azerbaijan in a period of heightened tension with Iran in 2001.[57]

The building of the BTC pipeline did not lessen, and may have increased, tension between Azerbaijan and Armenia. While some Western observers superficially favored a "peace pipeline" that would cross Armenia, the dynamics of the Nagorno-Karabakh conflict were far too risky for a large commercial project. In addition, domestic political opinion in Azerbaijan was too hostile to a route that rewarded a hostile state or invited sabotage. The construction of the $3 billion project gives BP and other investors a stake in a peaceful resolution of the conflict, especially since the pipeline runs just twelve kilometers north of the cease-fire line, but fears about Armenian attacks on the pipeline overstate Armenia's hostile intent and destructive capacity and understate the security measures that the pipeline investors have constructed.[58] Armenians are also aware that any sabotage of the pipeline by elements associated with Armenia would be an international diplomatic catastrophe.

Evaluation

As with security integration, political integration is largely lacking among Azerbaijan, Georgia, and Turkey. All three countries fundamentally lack a shared political orientation, with Georgia and Turkey sharing a predominantly pro-European perspective, while Azerbaijan has more complex motivations and would not hesitate to use its wealth to play Russia, the United States and the European Union, and Central Asian states against each other for its own benefit. As a corollary, Azerbaijan is less enthusiastic about joining the European Union or NATO than are Turkey and Georgia. Georgia's relations with Russia, from before the 2008 war, are profoundly troubled, while Azerbaijan maintains a kind of détente with Russia, and Turkey's relationship with Moscow tended to improve during the period in question.[59] The three countries have overlapping memberships in political organizations, with Azerbaijan a member of the Commonwealth of Independent States, Georgia suspended from membership, and Turkey a nonmember. Azerbaijan and Turkey are also members of a number of pan-Turkic and pan-Islamic political organizations that exclude Georgia. Nonetheless, the potential for political cooperation is high, as this section suggests: with common challenges in environmental protection of the Black and Caspian Seas, resource sharing, foreign occupation, and terrorism and the drug trade, Georgia, Azerbaijan, and Turkey are not without incentives for political cooperation in the future.

At the present, however, Deutsch's conditions for the existence of a security community are as yet unfulfilled. Deutsch wrote that a security community forms when political units become mutually interdependent and jointly produce system properties that they separately lack. While this may be occurring in other highly regulated sectors, at least in the South Caucasus it is not occurring in foreign policy, in governance, or in the rule of law—at least not yet. As discussed above for the security sphere and below for the political sphere, the countries share only the most superficial of common memberships in supranational organizations, and these organizations do not have supranational judicial or parliamentary powers. No common regional court yet exists, for instance. Indeed, the populations of the South Caucasus have no more than a very rudimentary sense of shared political identities. Certainly, overwhelming majorities in Azerbaijan and Georgia do not identify as European. When asked about identities, 71 percent of Azerbaijanis stated that they identified as Azerbaijani only, compared to 13 percent who would identify as Azerbaijani and European and 1 percent who would identify as European only. Statistics from Georgia were almost identical.[60] However, large majorities in Georgia favored strengthening ties with Azerbaijan: 44 percent answered it was "very important" and 38 percent answered "rather important," with identical numbers for Turkey and comparable numbers for Russia.[61] The intention for future political integration among Azerbaijan, Georgia, and Turkey may be there, but at the moment it is not any stronger than the wish for integration with Russia.

While high-profile political exchanges have occurred among all three countries, and while political rhetoric in both Georgia and Azerbaijan emphasizes certain shared values with Turkey (such as pan-Turkism in Azerbaijan and pro-European sentiment in Georgia), these shared values do not yet encompass all three countries to the exclusion of others. Deutsch wrote that contacts and transactions among politically relevant elites, such as diplomatic exchanges, election or cease-fire observer missions, common membership in and leadership of supranational political organizations, and a common foreign policy, could all help determine the shared political orientation of a multiunit polity such as Azerbaijan, Georgia, and Turkey. At the present, however, these contacts are too diffuse and contradictory to evidence a common "we-feeling."

Using the lenses of neofunctionalism and neoliberal institutionalism leads to substantially the same conclusions. Undoubtedly, regulatory integration and a standardization of customs, transport, and energy production have not spilled over into more robust political integration, but even using the more fundamental logic of neofunctionalism, it is not clear that political spillover is occurring either. Turkey sought to broaden its sphere of influence in Central Asia by emphasizing a common Turkic identity, and while it has had some success with Azerbaijan, it is not clear that this success is improving the prospects for pan-Turkism more broadly. Indeed, Turkey is becoming closer to Georgia than it is with its Turkic brethren.[62] The one exception to this is Armenia. As Turkey, Azerbaijan, and Georgia grow closer even in purely regulatory fields such as transport and energy, they do produce one major political consequence: they solidify Armenia's exclusion not only from regional markets but from world ones. This may be a form of limited political spillover, in which resource competition produces economic exclusion, which in turn produces political exclusion. However, it is unlikely that the three countries will be able to develop a common political union based solely on opposition to Armenia's occupation of Nagorno-Karabakh, or else, after twenty years, they would already have been successful. Georgia, while not sympathetic to the Armenian position in the conflict, benefits from being one of the only geographical connections between Armenia and the outside world.

Finally, although all three countries de facto exclude Armenia, even from a neoliberal institutional perspective it is not clear that membership in supranational organizations is necessarily drawing the three countries closer together. In fact, survey data show that there is little interest in forming a common identity based around supranational organizations such as the EU. Although organizations such as the Council of Europe promote common human rights, democracy, and rule-of-law norms in Azerbaijan, Georgia, and Turkey, the Council cannot make binding law. Even the common ratification of the European Convention on Human Rights and the acceptance of European Court of Human Rights jurisdiction by all three countries has not ensured compliance. Not all of the three countries have ratified the Council of Europe's terrorism, cybercrime, or democratic-reform conventions. Differing orientations toward the European Union and to the Commonwealth of Independent States likewise divide Azerbaijan, Georgia, and Turkey, and the three countries participate in a broad array of overlapping supranational political arrangements.[63] As with membership in the Council of Europe, the real impact of these organizations on the ground, considered in the aggregate, has been quite limited.

SOCIAL AND CULTURAL

The three countries in question are extraordinarily diverse for their size. The complexity of the relatively small region is defined by a mountainous landscape that has created a swirl of relatively insular linguistic and cultural minorities, even though the region's geography is actually quite varied.[64] Languages and religions differ substantially. Azerbaijan has a Shi'a Muslim cultural ancestry with close links to communities in northern Iran, and speaks a Turkic language. Georgia is historically Orthodox Christian, with a population that speaks a language unrelated to neighboring languages.[65] Although a natural link exists between Azerbaijan and Turkey due to their common linguistic background and at least traditional commitment to secular Islam, Georgia does not share in this common link. Both Azerbaijan and Georgia have experienced strong revivals of their traditional languages and cultures and "de-Russification" following the collapse of the Soviet Union. Whether this process of cultural and linguistic consolidation will draw these countries closer to Turkey is an open question.

Despite its position on the periphery of three historically important empires—the Ottoman, the Persian, and the Russian—the South Caucasus possessed a rugged terrain and sizable rural populations that prevented large-scale colonization by ethnic Russians, as had occurred elsewhere in the empire. The expansion of the British Empire and its sphere of influence into Central Asia in tandem with the decline of the Ottoman and Persian Empires effectively halted the Russian advance.[66] After World War I, the Allied powers supported a return to the status quo of Russian dominance after German-backed Ottoman expansionism was defeated in the region. The briefly independent states of Georgia, Azerbaijan, and Armenia were reincorporated into the Soviet Union.[67] Arguably, strong national identities only began to emerge in the three countries of the South Caucasus during the Soviet era, when a consolidation of national identity in the three countries homogenized their populations somewhat and caused many of the minority peoples to begin a process of assimilation.[68]

Although Azerbaijan and Iran possess religious and cultural affinities, relations between the two countries have never been friendly, particularly given Azerbaijan's strategic links with Turkey and the United States and the political activities of ethnic Azeris in Iran.[69] American-backed exclusion of Iran from the Azerbaijan Republic Oil Consortium, despite the support of Azerbaijan and the American oil companies involved in the project, ultimately worsened relations between Iran and Azerbaijan.[70] Iran has feared that the United States would seek to use Azerbaijan as a military base, and was skeptical of American military aid designed to be a warning to Iran. On the other hand, Iran was an important early mediator between Azerbaijan and Armenia over the Nagorno-Karabakh crisis, although its role in this position has declined over the succeeding years. The most serious tensions have flared over the Caspian Sea region, and an Iranian warship forced an Azerbaijani exploration vessel to leave the Alborz oil field in July 2000, sparking protest not only from the United States and Turkey, but from Russia as well.[71] These cool relations may be changing slowly over time. As Russia, Kazakhstan, and Azerbaijan have engaged in closer cooperation over Caspian Sea management, Iran has expressed a desire to become further involved in the multilateral efforts to partition the surface and the seabed.[72]

The secularism of Muslim Azeris is another factor that may naturally draw them toward Turkey instead of Iran, even though religious and clerical Azeris have reached high levels in Iran's religious elite. Azerbaijan's 1995 constitution sharply delineates a separation of church and state, and anticlerical sentiment is found not only in Azerbaijan but also among Azeri populations in northern Iran. Azerbaijani subjects under Russian imperial rule had access to secular education, and religious elites were heavily restricted in favor of secular intellectuals. A disproportionate number of secular Iranians have been Azeri and have found refuge in the Caucasus during periods of persecution.[73] The Azerbaijani population was subjected to six decades of antireligious propaganda during the Soviet period, as well as a ban on the teaching of Islam and disseminating religious literature. Despite the persistence of nominal or cultural adherence to Islam in much of Azerbaijan, these efforts by the Soviet state were remarkably successful.[74] In addition, because Azerbaijan is connected to both the Turkic and the Shi'a worlds, Azerbaijanis have played important roles in advocating ideologies that bridge Shi'a and Sunni Islam, such as pan-Islamic and pan-Turkic philosophies.[75]

With independence, Islam was no longer officially suppressed in Azerbaijan, and the country has been a member of the Organization of the Islamic Conference and the Islamic Development Bank since 1991. In a spirit of religious revival similar to the Central Asian republics, the country has seen the revival of old mosques and the building of new ones, as well as the establishment of religious societies and educational institutions. The liberal strata of the intelligentsia, the backbone of the ruling authorities, fear the rise of Shi'ite Islam and prefer the Turkish brand of moderate Islam. These elites have a strong European orientation and have developed notions of religious tolerance and modernization. However, the Azerbaijani brand of Islam was isolated under Soviet rule, and consequently the caliber of Islamic educational institutions is still debated.[76] Scholars in both Iran and Azerbaijan have focused on revitalizing the Azerbaijani language, including publishing works previously written during periods of oppression and helping to standardize the alphabet and vowel markers. Although Azerbaijani was not the primary language of many Azeris in Iran and Azerbaijan prior to independence, a concerted effort at expansion of the language helped to define Azeri post-independence identity.[77] The early 1990s in particular saw an outpouring of writing in the Azerbaijani language.

Technological changes have altered intergroup dynamics in Azerbaijan, Georgia, and Turkey, both encouraging and discouraging collective or complementary identities. In Azerbaijan, interaction with non-Azerbaijanis has usually augmented Azerbaijani identity, due to profound differences with the Russian population and lack of opportunity for social assimilation. By contrast, Azeri populations in Iran have assimilated into Persian culture and intermarried at higher rates. Television broadcasts from Turkey have had a striking impact on Azerbaijani identity, rapidly helping to make the Azerbaijani language more similar to Turkish and expanding awareness of a greater Turkic affinity.[78]

Unlike the largely homogeneous Armenia, Georgia and Azerbaijan are both very heterogeneous states and consequently vulnerable to outside manipulation in support of separatist movements and antigovernment forces.[79] Nor are the two countries without their own bilateral ethnic problems: the Azeri population of Marneuli in Georgia has agitated for incorporation into an independent Azerbaijan. Meanwhile the Georgian nationalist press has trumpeted the cause of ethnic Georgians living in northwest Azerbaijan, who have been subjected to oppression and linguistic assimilation.[80] In Azerbaijan, similarly, the Talysh, a Persian-speaking people to the south, have resented being submerged in Azeri Turkic culture, especially given the Soviet language-education policy that forced a choice between Azeri (the language of the autonomous republic) and Russian without regard to minority languages. The Lezghin, a Dagestani group to the north of Azerbaijan, were similar but their misgivings were less openly expressed.[81] Georgians abroad represent a very small diaspora, and the descendants of the Georgians who migrated to the West after Georgia's incorporation in the Soviet Union in 1921 have had little impact in the present era. The only semidiaspora is made up of Georgian Jews who have migrated to Israel but still possess some ties to the country.[82]

Since independence, in all countries of the South Caucasus and Central Asia there has been a trend away from Russian-language education and study abroad at Russian universities. A number of alternatives have gained currency, including study at universities in Turkey and the Middle East, as well as religious education, emphasizing cultural ties with these regions.[83] As a consequence, teachers and teaching materials used during the Soviet period were suddenly obsolete or irrelevant, and in many cases, funds or professional capacity in the new languages and scripts have lagged behind. Coupled with endemic corruption in the education system and physical deterioration of infrastructure, the crisis in educational policy in a number of post-Soviet independent states has led to brain drains to the West and disillusionment and political activism among youth. The deterioration of secular state-sponsored educational systems in the Caucasus and Central Asia has led to an erosion of social cohesion, resulting in intolerance and xenophobia, religious extremism, and lack of respect for state authorities.[84]

Evaluation

From a sociocultural perspective, Azerbaijan, Turkey, and Georgia share some important similarities, but are not integrating among themselves in a single regional unit apart from their neighbors. While all three countries are undoubtedly more globalized than at any point in their histories, each possesses complex relationships with societies and cultures on their peripheries. Georgia and Azerbaijan are in the process of developing modern national languages, literatures, and educational systems already possessed by Turkey, and both countries are undergoing a transition away from the Russian model. Beyond this, the similarities end. The Azeri and Turkish languages may become more similar as they are globalized due to the telecom revolution in the region, but the Georgian language is highly distinct and does not even possess a common script with the latter languages. Likewise, Azerbaijan and Turkey are a model for secular Islam, particularly as contrasted to Iran, even though Azerbaijan is a Shi'a country while Turkey is a Sunni one. However, Georgia is an Orthodox Christian country that shares many religious traditions with Russia. While both Georgia and Azerbaijan are in a post-independence nationalist period in which arts, literatures, and scientific and university elites are helping to develop a national identity and standardize the national language, Turkey is far more developed in this regard and undoubtedly has a unitary and hegemonic (if often contested) national culture. In addition, it must be remembered that Turkey dominates the three-country grouping with a population of seventy-three million people, many times larger than Azerbaijan (nine million) or Georgia (four and a half million).

Some commentators argue that at the broadest level, all three countries seem to be moving toward a more European or Western sociocultural orientation, although Azerbaijan in particular seems less unidirectional or consistent than the other two in this regard, and certainly does not share Georgia's or Turkey's enthusiasm for joining the European Union or Georgia's enthusiasm for joining NATO.[85] This accords with Keohane's neoliberal institutionalist theory, because this "Europeanization" of identity would in part be facilitated by European supranational actors working in the region. But, there is some evidence to suggest that this definition of Europeanization and integration do not necessarily go hand in hand. European-sponsored institutional change has helped to preserve existing elites in power rather than change them, and incumbent regimes have selectively used anticorruption and democratization efforts against their political opponents.[86] Existing elites often have control over the strategic resources that the European Union seeks (gas, oil, transport corridors), and have a natural advantage in cultivating European support. This is the ambivalence at the heart of neoliberal institutionalism: should supranational agencies, or other transnational extra-regional actors, funnel resources through official state channels, they have the consequence of strengthening the existing state; should supranational agencies funnel resources around official state channels, they weaken the state. It may be the case—although it is beyond the scope of this study—that Azerbaijan, Georgia, and Turkey may be collectively moving toward a more pro-European worldview. However, should the legitimacy of their regimes come into question, this pro-European tendency could be reversed.

Although sociocultural integration is central to security communities theory, Azerbaijan, Georgia, and Turkey also fail to fulfill the prerequisites for neofunctional or neoliberal institutional cooperation. All three countries are members of regional scientific, educational, and cultural organizations and participate in scientific and academic transfers, but they do not participate in these activities to the exclusion of other countries such that an integration process could be said to exist. Azerbaijan and Turkey are members of the Turkic Council in Istanbul, formed in 2009, and the Joint Administration of Turkic Arts and Culture in Ankara, formed in 1993, but Georgia is not a member of either organization. Likewise, Azerbaijan and Turkey are members of the Organization of Islamic Cooperation, but again Georgia is not. Georgia and Turkey do have significant academic exchange, just as Turkey does with Azerbaijan, and the International Black Sea University (IBSU) was founded in Georgia in 1995 by Turkish companies in an attempt to bolster Georgian-Turkish relations. Instruction is in English. Georgia maintains a pro-Western orientation and has even established academic programs on American studies at IBSU.[87] However, from a neofunctionalist perspective, this scientific and academic cooperation does not seem to have produced significant regulatory changes, such as lowering the barriers for student visas, although all three, as members of the Council of Europe, have signed the Lisbon Convention regulating degree qualifications and periods of study.

Short of the vague and diffuse process of Europeanization (especially as to Georgia and Turkey), societies and cultures increasingly overlap among the three countries, but are not yet interdependent. The swirl of linguistic, religious, and ethnic groups among Azerbaijan, Georgia and Turkey still remains insular, even as the newer nation-states try to consolidate their linguistic and cultural identities. The telecommunications revolution, discussed in more detail below, may make some impact, but in every case the integrating units can only count to two: the Turkification of the national language (Turkey and Azerbaijan); the secularization of Islam (Turkey and Azerbaijan); the post-Soviet consolidation of new national cultures (Azerbaijan and Georgia); and affection for European civilization (Turkey and Georgia). A sociocultural core common to all three countries, such as a common language or religion, or even a common commitment to principles of governance and the rule of law, has not yet formed among Azerbaijan, Georgia, and Turkey.

ECONOMICS AND TRADE

Economic integration in the South Caucasus, like political and sociocultural integration, is uneven and multidirectional. Azerbaijan's role in world trade is dominated almost exclusively by fossil fuels, accounting for about 96 percent of the country's exports.[88] Given the country's distance from Europe and its generally less enthusiastic European partnership compared to Georgia, it may be surprising that the European Union is by far the country's largest trading partner, receiving nearly half of Azerbaijani exports and accounting for about 25 percent of the country's imports, more than Russia. Georgia, by contrast, has a far more diverse portfolio, with manufactured goods accounting for more than half of Georgian exports, and the rest evenly divided between agricultural goods and fuels and mining products. Georgia's main export partners are the European Union (18.5 percent), Azerbaijan (15.4 percent), and Turkey (13.6 percent), which may be predictable given the country's strategic location. Similarly, the overwhelming majority of Georgian imports arrive from the European Union, Turkey, Ukraine, and Azerbaijan. Finally Turkey's trade is primarily with the European Union, the Russian Federation, China, and the United States.[89] These raw numbers are misleading, of course, as Turkey's trade volume is, in aggregate monetary terms, ten times the trade volume of Azerbaijan, and Azerbaijan's trade volume is at least five or six times that of Georgia. Given Georgia's geographic proximity to Turkey and Azerbaijan and its role as a conduit for the high-value resources coming from the Caspian basin, it is little surprise that it is the South Caucasus state that is the most economically well integrated in the region, but also that its trade constitutes a fairly minor proportion of the total.

Trade patterns among the countries of the South Caucasus have greatly changed since independence. Improved relations between Turkey and Russia have led to an expansion of growth in the South Caucasus, halted only briefly by hostilities such as the 2008 war between Russia and Georgia. Since the beginning of the twenty-first century, trade volume among Azerbaijan, Georgia, and Turkey has been steadily increasing: as between Georgia and Turkey, from $241 million in 2002 to over $830 million in 2007; as between Azerbaijan and Turkey, $296 million in 2003 to $1.2 billion in 2007; and as between Azerbaijan and Georgia, $76 million in 2000 to $411 million in 2006.[90] Since 2006, Russia has become Turkey's largest source of imports, with natural gas and crude oil figuring prominently: about half of Turkey's crude oil and 65 percent of its natural gas come from Russia. The trade balance between the two countries heavily favors Russia and was expected to reach $38 billion in 2008, up from $27 billion the prior year.[91] By contrast, Turkey's trade with Georgia and Azerbaijan is much smaller. For its part, Azerbaijan has had unprecedented economic growth since the mid-1990s. While GDP growth was negative in 1995 and only 1 percent in 1996, it soared to 26.4 percent in 2005 and is expected to remain positive. The trade account has also been positive since 2000 due to increasing exports of oil and decreasing imports of machinery and inputs for the oil and gas sector. Government spending in the form of energy subsidies, pensions, and wages has also increased, boosting private consumption.[92]

When compared to the greater political integration in postwar Europe due to free trade and open borders, the South Caucasus countries are painfully estranged from one another. The semi-isolation of Georgia from either Azerbaijan or Armenia is most obvious at the borders: travel can take hours; commercial travel is subject to high tariffs and substantial bribes; traffic police often target cars with foreign license plates for harassment and bribes.[93] Significant legitimate trade between Georgia and its neighbors is limited to timber, wheat, and energy products. The political economy of economic isolation in the South Caucasus is significant and deeply rooted; it is not simply a matter of corrupt local officials sabotaging the entrepreneurial plans of political elites. This structural political tendency to favor short-term political interests over long-term economic development provides strong disincentives to joint ventures, let alone a Caucasus common market.[94] The political vulnerabilities of porous borders and corruption are not limited to organized crime and black-market trade. In 2001, the presence of several dozen mujahedeen, several with ties to Al-Qaeda, made Georgia a haven for international terrorism due to the weakness and corruption of security forces. In Pankisi, a disastrous expedition by Chechen commander Ruslan Gelayev to the Kodori gorge in Abkhazia further undermined the peace process and Georgian statehood.[95]

Remittances are important for all three South Caucasus states, particularly Armenia, which has a small resources base and a moribund manufacturing sector. The annual income of one million Azerbaijani citizens in Russia ranges between $700 million and $1 billion, while more than 500,000 Georgian workers in Russia contributed one quarter of Georgia's GDP before the 2008 war. Azerbaijan and Georgia stand to benefit from Turkey's economic links to the West, particularly as Turkey is a transport hub for a region that has suffered from closed markets due to political instability, a dearth of transit routes, and high tariffs.[96] Turkish-Russian cooperation expanded due to Turkish exasperation with US policy in Iraq, and a rapprochement between the two countries included high-profile meetings between the presidents of Turkey and Russia, a new gas pipeline under the Black Sea, and negotiations on future energy deals. Turkish construction and consumer goods companies have major ventures in Russia, and Turkey is the top destination for Russian tourists, who numbered over 1.7 million in 2004. By contrast, Turkish-Georgian relations since the Rose Revolution have stagnated over Georgia's contentious relationship with Russia and close relationship with the United States.[97] As Turkish-Russian relations improve, economic intertwinement between Turkey and Georgia may slow.

As early as the mid-1990s, Western countries have viewed Georgia as the most progressive post-Soviet country outside the Baltic states, and double-digit growth between 1996 and 2000 and the influence of young reformers such as Zurab Zhvania, the Speaker of Parliament, provided at least a superficially positive picture. However, Georgia remained a weak state captive to political factions, with corruption in important agencies such as the Interior Ministry and unholy alliances with warlords and rebel factions. The final three years of Shevardnadze's rule, until his resignation in 2003, saw sharp erosion in public legitimacy due to corruption and a Chechen refugee crisis.[98] Election irregularities in November 2003 led to massive public outpouring of support for Saakashvili's National Movement, who exit polling had shown had won first place in the elections. The Rose Revolution proved to be a sea change for economic integration in the region. Despite the more authoritarian aspects of Saakashvili's presidency, he succeeded in overhauling the customs service and tax police, simplifying tax rates, slashing red tape, and forcing companies to declare their incomes. New roads, hospitals, and schools were built; wages and pensions increased; and foreign direct investment was also aided by a public relations campaign. Saakashvili also ousted the corrupt regime in Ajaria and reintegrated the autonomous province back into the Georgian state.[99]

At least until the Rose Revolution, economic integration in the South Caucasus was hampered by the informal economy. In Georgia, widespread poverty, an informally privatized economy, and weak and corrupt law enforcement made the black market all-encompassing and have endangered attempts to build a properly funded and functioning state. The informally organized private interests who control timber in northern Armenia and western Georgia, hazelnuts and cigarettes in Ajaria, many kinds of petroleum products in Azerbaijan, or scrap metal all over Georgia have much to lose from free trade.[100] Upon taking office, Saakashvili pledged to tackle corruption and border issues and showed commitment by arresting corrupt figures in the former Shevardnadze regime, conducting operations against smuggling in Ajaria and on the Georgia-Azerbaijan border, and attempting to clean up the customs services.[101]

In addition to these economic reforms, the democratization process in Georgia greatly changed the propensity for corruption and black market operations since the Rose Revolution, the first widespread democratic movement in a former Soviet state outside the Baltics. When Saakashvili took power, he faced daunting economic problems due to massive corruption and economic mismanagement.[102] Under Saakashvili, budget revenues grew from $432 million in 2003 to nearly $3 billion in 2008, and both services such as security and energy distribution and infrastructure projects flourished. However, although Georgia succeeded in strengthening state institutions, increasing their nationwide presence and projecting power outside the capital city, all preventing the political crisis from escalating to the degree that it did in the early 1990s, the transformation alienated local autonomous political leaders and worsened relations with Russia.[103] Initial Turkish opportunism has been replaced with increasing direct Turkish investment and trade with Georgia. Cooperation leading to the BTC project has contributed to the standardization of certain economic regulations, and investors additionally required certain business, human rights, and environmental standards to be in place before committing money to the project. A transnational business such as the pipeline required standardized accounting, wage, and other formal business practices. Additionally, Turkey has replaced Russia as Georgia's largest trading partner.[104]

For most of the independence period, the porousness of borders in the Caspian region has caused undesirable forms of trade to expand with the desirable ones, including narcotics trafficking.[105] Drug trafficking distorts healthy economic forces, and dealers in Central Asia and the Caucasus have found ready markets. Drugs seep into the Caucasus from Iran. Georgia is particularly hard hit by the international drug trade, particularly heroin and raw opium bound for the European market from Afghanistan via Iran, or from Turkmenistan over the Caspian Sea.[106] Russian troops stationed in Armenia discovered a drug channel in 1997 and in 1998 Georgian police confiscated opium grown in Pakistan; also that year, more Iranians were arrested on drug charges than citizens of any other country. Security forces in the Caspian region stage hundreds of drug busts per year, but these efforts are insufficient to halt the narcotics trade.[107]

Azerbaijan and Georgia are not strategic competitors; rather, they are emerging as strategic partners. Azerbaijan has a powerful resource in its energy supplies, and Georgia has one in its transit value and location. As a consequence, the absence of economic rivalry between the two countries belies their commitment to capitalism and market reform; rather, the partnership has interregional competitors, and the association of the two countries will prove attractive to investment in the long term.[108]

Evaluation

Economic integration in the South Caucasus has achieved more success than political or security integration, and regional trade patterns have undoubtedly shifted away from Russia and toward Turkey and the European Union over the past twenty years. The fragility of this integration, however, was underscored by the Russia-Georgia war of 2008, which reminded potential trade partners that the energy and transit route through Azerbaijan, Georgia, and Turkey is exposed and vulnerable to Russian influence. The region as a whole has generally shown strong economic growth. Azerbaijan's real GDP grew at least 200 percent between 2005 and 2010, and Turkey's much larger economy also registered consistent overall growth from the mid-1990s to 2008. While Georgia's real GDP grew somewhat between 2005 and 2010, it actually declined in 2009. As is fitting of a transit country, Georgia's trade in commercial services, including transport, is robust and rivals Azerbaijan's despite Azerbaijan's larger economy and population. Even this sector, however, suffered considerably in 2009 after the Russia-Georgia war.[109] Additionally, as with sociocultural and political integration, the extent to which economic integration has occurred bears some evidence of cultivation by supranational institutions such as the World Trade Organization (of which Georgia and Turkey are members) and the European Union. In this process, Georgia is much further ahead toward negotiations over entering into a comprehensive free-trade agreement with the European Union. Azerbaijan cannot yet benefit since WTO membership is a prerequisite for these agreements and Azerbaijan's membership in the WTO is still pending.[110] All three countries benefit from the EU's generalized system of preferences for the EU market.[111]

Despite efforts at economic integration among Azerbaijan, Georgia, and Turkey, the process is as yet incomplete. From a neoliberal institutionalist perspective, the three countries possess conflicting memberships in international organizations and do not necessarily share the same economic integration strategies, particularly as to how to manage the relationship with Russia. Turkey and Azerbaijan are members of the Economic Cooperation Organization (ECO), founded in 1985 by Turkey, Iran, and Pakistan, and Turkey heads the ECO Trade and Development Bank based in Istanbul. Azerbaijan is a member of the Commonwealth of

Independent States, while Georgia is suspended from membership and Turkey is not a member; none are members of the Eurasian Economic Community. Georgia and Azerbaijan are members of the GUAM (Georgia, Ukraine, Azerbaijan, Moldova) Organization for Democracy and Economic Development, and Turkey is an observer. The organization was created in 2001 and is seen as a way of countering Russian influence, in investments as well as politics, in the Black Sea region. Despite initial setbacks, the organization has had limited success in cooperation on issues of electricity and power, transport, telecommunications, and organized crime and the drug trade. GUAM is an apt analogy for economic integration among Azerbaijan, Georgia, and Turkey generally: limited success in highly regulated and high-value fields such as energy and transport, an ambiguous role for Russia, and only partially shared economic goals among the member states.[112]

ENERGY

Baku has a long history of involvement in the production of commercial oil, and was even the world's energy capital in the late nineteenth century when the region produced more oil than the United States.[113] From 1871, when the Russian government permitted the drilling of the first private oil wells, the Baku region has been the center of an enormous industry, transformed in a single generation "from a forgotten desert citadel into a modern metropolis."[114] Almost immediately the problem of transport surfaced. While oil was transported for centuries with camels and yaks and, when not stormy and dangerous, across the Caspian Sea to Russia, the opening of the first railway from Baku to Batumi, Georgia, in 1883 opened Azerbaijani oil to the Black Sea. In 1906, a kerosene pipeline opened between Baku and Batumi.[115] Azerbaijan's oil production stalled after it joined the Soviet Union in 1920, peaked during World War II when it accounted for about 70 percent of the USSR's production, and relapsed again in the postwar years after Russia shifted its attention to the Volga-Ural region. When the USSR disintegrated, Azerbaijan was producing about 60 percent of Soviet extraction machinery and spare parts, but only about 2 percent of its oil.[116]

Although Azerbaijan became the first former Soviet republic whose borders threatened to change as a result of warfare when it became independent, a broad combination of factors made investment attractive despite some concerns over the country's stability. The Caspian region is generally attractive to Western oil companies due to proven reserves of nearly twenty billion barrels, with the Azerbaijani section of the Caspian Sea alone conservatively estimated at four billion barrels.[117] The prospect of joining the vast Caspian oil and gas reserves with those of Turkmenistan and Central Asia in a unified pipeline network that was open to

European markets was particularly attractive to Western oil companies. In addition, the opening up of the Azerbaijani economy coincided with both the Clinton administration's desire to reduce dependence on Middle Eastern oil and the transfer of the Russian state monopoly on oil to private oligopoly with limited participation by Western oil companies. Unlike in the Persian Gulf states, Western oil companies have secured a major foothold in the Caspian region.[118] Oil from Azerbaijan profited significantly from the oil embargo on Iraq, although in general OPEC states such as Iraq possess a greater proportion of the world's proven oil reserves while non-OPEC states such as Azerbaijan form a greater portion of the global market. As a consequence, non-OPEC output will begin to decline sooner than OPEC output.[119]

Among the South Caucasus countries, Azerbaijan is undoubtedly the leader in the energy sector. Since 1992, Azerbaijan has consistently prioritized development of its offshore fields in increasing the country's oil resources through the State Oil Company of the Azerbaijan Republic (SOCAR), despite regime changes since independence. Unlike early onshore exploration and development contracts, the offshore agreements were framed as production-sharing agreements (PSAs), which boast a significant advantage over other contractual arrangements: ratification by the Azerbaijani parliament and consequently the force of law.[120] This provides increased levels of contract security and insulation from notoriously invasive changes in Azerbaijani domestic law. PSAs are better than traditional tax-and-royalty schemes as a mechanism for providing the Azerbaijani state with its share of the profits without endangering the private investment, and they allow participants who invest capital and operating costs a share of crude production, with the remainder of an oil field's output divided between the state and its foreign partners.

Aliyev helped secure Azerbaijan's fate by securing strong multinational commercial interests and using the revenue to fortify himself and Azerbaijan against foreign and domestic enemies. BP's signing of massive, multidecade offshore contracts with the government of Azerbaijan and other foreign partners helped to draw major Western governments strategically closer. The wide-ranging offshore PSAs ensured hard-currency revenues for the economy over the long term as well as substantial and immediate payments to the Azerbaijani government.[121] SOCAR has a virtual monopoly over the Azerbaijani oil sector, including relations with foreign companies, negotiation and implementation of projects, and international marketing. A web of informal networks linked top officials in SOCAR with Aliyev to create a nucleus of decision-making power outside formal government hierarchies, which gave Aliyev great control over the distribution of revenues. In conflicts with other state regulatory bodies, including those that control investment, privatization, or customs, SOCAR has almost unfettered influence and is especially resistant to administrative restructurings and leadership changes.[122] Seen as a "one-stop shop" by foreign investors, the sheer bureaucratic power of SOCAR and its near-total dominance over the oil sector is a mixed blessing, as SOCAR contracts are negotiated outside official channels and have the force of law with little oversight by parliament. While this may be profitable for an oil partner, it can also expose foreign companies to irregular legal, tax, and regulatory scrutiny by removing all effective means of oversight.[123]

By 2010, the Turkish economy was expected to depend on energy imports for about 90 percent of its needs.[124] Unlike Azerbaijan, Turkish energy policy is decentralized among several overlapping government agencies, which has led to institutional rivalry and duplicity, particularly with clashes between the Ministry of Foreign Affairs and the Ministry of Energy. Energy issues have both technical aspects and political aspects, and attempts to resolve some of the rivalry have led to, for instance, the creation of the Baku-Ceyhan Working Group, composed of various energy officials and experts from both agencies.[125] Given the increasing closeness of Georgian and Azerbaijani relations, Georgia's territory was highly suitable for the transit of oil from Azerbaijan, which led to the opening of a pipeline from Baku to an oil terminal at the Black Sea port of Supsa, Georgia, in April 1999.[126]

The BTC pipeline had evident political benefits for Azerbaijan and Georgia. For Georgians, the pipeline had an overwhelmingly political character and has come to symbolize Georgian independence as a state, while Azerbaijan benefited from the cooperation of a politically weak transit country such as Georgia in place of dependence on Iran or Russia. The BTC pipeline now has near-universal backing in Georgia and Azerbaijan, and no significant political voices have been raised in opposition in either country. The notion that the pipeline will help Georgia become an east-west transit corridor independent of Russia has almost universal acceptance.[127] Despite initial skepticism from oil companies at its inception, the viability of the Baku-Tbilisi-Ceyhan pipeline was boosted greatly by the additional oil discoveries in the Kashagan offshore oil field in Kazakhstan in 2000, which will require additional outlets for oil transit.[128] Kazakhstan, with American urging, came to officially support construction of the Baku-Tbilisi-Ceyhan pipeline.

The BTC pipeline may have helped provoke Russia into an antagonistic posture. The project certainly polarized the region between an Azerbaijan-Georgia-Turkey axis, supported by Western backers, and a Russia-Armenia-Iran axis on the other.[129] Armenia has turned to Russia for its energy needs and signed a deal with Iran to construct an Iran-Armenia gas pipeline, completed in 2006, with Russian gas giant Gazprom as majority stakeholder.[130] In Russia, a major difference exists between the oil and gas sectors. Oil is handled by a number of semiprivatized companies. Gas is controlled by the wholly state-controlled giant Gazprom, a powerful instrument for the government to impose its will on natural gas production and export, a power that has been used and occasionally abused.[131] In the mid-1990s, Russia placed political pressure on Kazakhstan to sign large, long-term delivery contracts with Gazprom in an effort to isolate Azerbaijan and prevent Turkey from gaining ground in oil and gas export lines. However, the sheer size of oil and gas reserves in the Caspian forced Russia to make occasional concessions, such as allowing Turkmenistan to deliver natural gas to Turkey via Iran.[132]

The exclusion of Iran from American-backed oil-transport schemes has ultimately increased in importance a Caspian Sea pipeline route, despite the earthquake-prone nature of the region and the existence of other environmental hazards. Undoubtedly, the most attractive oil-transit routes could well be through Iran to the Persian Gulf, but sanctions on the country prevent development of perhaps the most cost-effective and profitable arrangement.[133] The Iranian sanctions are a gift to Georgia, which became, by default, the only avenue for Caspian oil to avoid Iran and Armenia. Even the ports of Poti and Batumi, under the control of the Russian navy, were doubtful. As a consequence, in late 1998 Georgia built a terminal at the obsolete tourist resort of Supsa.[134] The facilities at the Ceyhan terminal are superior to those at Novorossiysk. Ceyhan is able to handle up to 120 million tons per year while Novorossiysk only has a loading capacity of 36 to 38 mt/y, with only another 30 possible after a second terminal is built, although this has already been allocated to the Tengiz oil field in Kazakhstan.[135]

The natural gas crisis of 2006, when Russia shut down supplies to Ukraine in the middle of winter, revealed just how precarious Europe's dependence on Russia as supplier of natural gas was, and the Russia-Georgia war in 2008 and a second Russia-Ukraine gas crisis in January 2009 reinforced these perceptions.[136] Azerbaijan became a world gas player when BP discovered natural gas in Shah Deniz in 1994, and the South Caucasus (Baku-Tbilisi-Erzurum) gas pipeline was constructed alongside the BTC oil pipeline for most of its route. While the country possesses about 1.2 trillion cubic meters of natural gas, about one-sixth of Turkmenistan's supply, Azerbaijan benefits from a Western-run gas pipeline already connecting the Caspian Sea to Western markets.[137] However, efforts to connect the gas pipeline at Erzurum to Central Europe have met with competition from Russia. Project plans for a Western-backed pipeline from Erzurum to Austria, named Nabucco, have sparked a Russian alternative to run across the Black Sea to Bulgaria, which may produce oversupply for Central and Eastern Europe's demand and render both pipelines economically infeasible.[138] Azerbaijan has the potential to be a major gas producer for the next twenty years, and to be the conduit for an even bigger world gas power, Turkmenistan. However, given the current political climate and the instabilities of the Caspian Sea region, it is unlikely that a natural gas pipeline will be built connecting Turkmenistan to Azerbaijan, Georgia, and Turkey in the near future.[139]

By 2010, Azerbaijan was, by far, a much wealthier and more confident country than it was fifteen years prior, with GDP growing spectacularly from $1.3 billion in 1994 to $25 billion in 2008. However, current projections are that production from the ACG fields that fills the Baku-Ceyhan pipeline will start to decline steadily during this decade, forcing the country to begin contemplating a post-oil economy. Under heavy international encouragement, Azerbaijan has set up an oil fund to help the population deal with a decline in revenues, but diversification of the economy is especially difficult in countries that have a high dependence on oil wealth, and the industry is notoriously frugal in creating employment for local populations.[140]

Since 1999, the State Oil Fund of Azerbaijan (SOFAZ) has been the main mechanism for managing the country's oil wealth, with assets that reached $1.5 billion in 2006 and are expected to exceed $50 billion in the next eighteen years. A portion of the assets are used for investment and social projects, including the construction of housing for refugees and internally displaced persons, which accounted for about $40 million of the $373 million that the fund spent in early 2006. In addition to transfers from SOFAZ, tax payments to the state oil company SOCAR accrue to the state budget, totaling $320 million in 2002.[141] Azerbaijan has received some praise for its redistribution of oil profits in mitigating the losses of vulnerable population segments in order to avoid social tensions, increasing both the amount of state benefits and the number of beneficiaries.[142]

Azerbaijan has not been wholly separate from Iran, another potential outlet for Caspian Sea oil and gas. In May 1995, Azerbaijan announced that crude oil produced by the Azerbaijan International Operating Company would be exported to the Persian Gulf in an Iranian-Azerbaijani oil swap, leaving open the possibility of further shipments of oil through Iran to the world market. Moreover, that year Iran also formed an exploration company with Russia and Azerbaijan to prospect for oil in the Caspian Sea, and Azerbaijan and Iran agreed to drill jointly in Iranian territory according to at least one of the contracts. Baku and Tehran also cooperate on hydro projects on the Alas River on the Iranian-Azerbaijani border.[143] The Caspian Sea basin is rich enough in oil and gas to serve many routes of transport for the distant future. Apart from a network of Soviet-era pipelines serving Russian and Kazakh oil fields, a pipeline also passes from Baku via Grozny, Chechnya, to the tanker terminal of Novorossiysk on Russia's Black Sea coast.[144]

Evaluation

Undoubtedly, some integration has occurred in the energy sector among Azerbaijan, Georgia, and Turkey, in contrast to the security, political, and sociocultural spheres. As noted above, energy is a high-value resource in a well-regulated field, and the actors who stand to make a profit from energy exploration, exploitation, and delivery to world markets have incentives to reduce transaction costs by removing traditional barriers erected by the nation-state such as customs and tariffs. Pipeline finance and construction requires comprehensive government commitments and contracts that are immune from political pressure. In limited ways, pipeline construction also places pressure on regulatory reform efforts to prevent corruption in regards to concessions and licenses, promote transparency in the distribution of oil revenues, reduce trade barriers, and increase pipeline security. This is not spillover in the traditional neofunctionalist sense, and many of these reforms do not extend out of the energy sector. Nonetheless, empirical evidence strongly suggests that a process of integration is actually occurring among Azerbaijan, Georgia, and Turkey.

Energy integration may reveal some of the limitations of each of the integration theories explored in the prior chapter. Undoubtedly, harmonization of energy regulatory regimes is occurring among Azerbaijan, Georgia, and Turkey to the exclusion of their neighbors, forming a distinctive community, at least in that sphere. As noted, this has not led to spillover in the classical sense, and in some ways the reverse may be true: energy policies in each of the three countries may be treated as sui generis cases, separate from other regulatory policies. This is particularly true in Azerbaijan, where decisions about energy resources are made outside the formal bureaucratic structure. As energy policies grow more harmonized among the three countries, energy policy itself is becoming further removed from ordinary policy making and legislation. Couched in special agreements and contractual relationships, energy policy is treated differently from other forms of policy. This not only limits the potential for spillover into other sectors, but subverts the spillover process. As energy policies become more harmonized, they become more distinct. Neofunctionalist theory cannot completely explain this phenomenon. However, neofunctionalism's emphasis on reducing transaction costs is relevant to energy sector integration.

Security communities theory similarly does not fully account for energy integration among the three states. Decisions about energy policy are made far from the public sphere, and multinational energy companies are particularly notorious for their opaque deal making, which lends itself to corruption or at least over personalization of contacts. Transactions or communications among populations across borders seems to have little relevance to the process of energy integration, although public perceptions and expectations become self-reinforcing as energy integration takes place. Rather, energy integration by its nature is a more top-down phenomenon, certainly more so than transport, as explained in the next section. Transnational business interests and international financial institutions play a considerable role in facilitating energy integration through resource exploitation and pipeline planning, construction, use, and security. As the neoliberal institutional model would predict, transnational extra-regional actors have placed pressure on the three national governments to change their patterns of behavior, reduce transaction costs, ensure compliance, and maintain creditworthiness and reputational value. This has helped to facilitate harmonization among the three countries without yet at least causing spillover in other sectors.

TRANSPORT

Transport is one area in which integration among Azerbaijan, Georgia, and Turkey, to the exclusion of their neighbors, has led to integration of national sectors into a subregional whole. In many ways, this was dictated by resource constraints and environmental realities. The Bosporus Strait, which connects the Black and Mediterranean Seas, dictated Turkish policy toward pipeline construction and tanker shipping from the South Caucasus to world markets. Increased shipping volume through the Bosporus, such as would occur if Caspian oil were transported in greater quantities to the Russian port of Novorossiysk, could have grave environmental consequences. Turkey sought to curb the prospect of an oil spill with a vessel-tracking system to facilitate safe passage through the straits, but it would not eliminate the risk entirely. Some nineteen miles long and a mere seven hundred meters wide at its narrowest point, the Bosporus is one of the most difficult waterways in the world to navigate, with tankers having to change their course at least twelve times along the route. Although 45,000 vessels pass each year, accidents are relatively frequent.[145] The Montreux Convention, adopted in 1936, requires free passage of the straits and severely constrains the ability of Turkey to monitor safety of passage. The year-round, warm-water Turkish port of Ceyhan on the Mediterranean can handle many times the capacity of the Russian port of Novorossiysk, and without endangerment to the Bosporus Strait.[146] If all Caspian oil output by 2002 were to travel through the Bosporus by tanker, it would require closing the straits to all other maritime activity for 360 out of 365 days of the year. Today, many experts believe that the Turkish political leadership may have overstated the issue of environmental fragility in the Bosporus to place pressure on Russia to accept a BTC pipeline.[147]

Just as the BTC oil pipeline and the BTE gas pipeline connect the energy infrastructure of Armenia, Azerbaijan, and Turkey, a new railroad between Baku, Azerbaijan, Tbilisi, Georgia, and Kars, Turkey, is "another step in the definition of the South Caucasus" and another component in the emergence of the region as a major transit corridor between Europe and Asia.[148] Following the closure of an earlier route that traversed Armenia, attempts at connecting the three cities by rail in a single line floundered until the presidents of Azerbaijan, Georgia, and Turkey floated the project idea at the inauguration of the BTC pipeline project in May 2005.[149] The railroad is expected to open in late 2011, with a length of 826 kilometers and an initial capacity of 1 million passengers and 6.5 million tons of freight.[150] This will increase to 3 million passengers and over 15 million tons of freight.[151] Baku has been connected by railway since 1883 and by pipeline since 1907 to the Black Sea port of Batumi, Georgia, the main outlet for oil exports in the early years.[152]

The implementation of transport infrastructure among Azerbaijan, Georgia, and Turkey is highly linked to the further informal integration of these three states. The building of the BTK railroad may enable the three countries to "vary the basis of their political cooperation" beyond energy transit and ancillary processes, "orientating it towards goods transport, but also towards higher circulation of people between the three states."[153] Because of political friction over the BTK railroad's bypass of Armenia, Azerbaijan has taken a prominent role in financing the railroad out of the country's energy revenues, the first time a transnational project is proceeding without the involvement of third countries or international organizations.[154]

Turkish President Demirel was warmly received in Tbilisi in January 2000, where he held discussions on possible construction of a Kars-Tbilisi railway, rehabilitation of the Batumi airport, and the opening of a fourth border gate between the two countries.[155] The railroad will be the first direct rail link connecting Azerbaijan and Turkey, and given that the bulk of the railroad's length is in Azerbaijan, the country will benefit the most in tariff revenues.[156] The more prominent role taken by Azerbaijan as financier of the BTK railroad has likely tempered the desire of Georgia and Turkey to reopen the Kars-Gyumri-Tbilisi railroad, which runs through Armenia, and will likely further encourage the isolation of Armenia within the South Caucasus.[157] Because Armenia's borders with Azerbaijan and Turkey are closed, Armenia is entirely dependent on Georgia for foreign trade with Russia or Europe, despite high Georgian tariffs.[158] Consequently, Armenia's economy suffers when the Georgian economy suffers, as it did during the Russia-Georgia war in 2008 that led to the bombing of the Gori-Tbilisi railroad and the blockade of Georgian ports on the Black Sea.[159] The 2008 bombing also damaged Georgian ports that are fed by the rail line, particularly after the bombing of Poti, which, along with Batumi, is the key port facility for the entire region. The Poti bombing also briefly closed the Georgian port of Kulevi, which is owned by the Azerbaijan state-owned oil company SOCAR, underscoring the risks of integration when relations with Russia sour.[160]

The idea of building a railway connecting the Black and Caspian Seas was first proposed in the 1830s upon the realization that a railway from Poti on the coast to Tbilisi in the interior could carry a great deal of trade between northern Iran and Europe, and only half as long as a railway through Turkey. Construction of the first trans-Caucasus line started in

1865 and the first passenger train from Poti arrived at Tbilisi on October 10, 1872, covering the journey in fifteen hours. By 1890, an extension through western Georgia, including the double-track Tsipa tunnel, was completed. In 1883, the line was opened between Tbilisi and Baku to carry oil from the Caspian oil fields to Batumi, and the last major line between Georgia and Armenia opened in 1899. As Georgian industry rapidly developed between 1924 and 1940, new branch lines of Georgian railways opened throughout the country, with a connection to the Russian railway network opening in 1946. Completion of the network between Baku, Tbilisi, and Batumi had a positive effect on the economies of Georgia and the South Caucasus. In the 1870s, Batumi was a major exporter of wool, cotton, lumber, and manganese. In addition, a tourist line was built in 1894 to Borjomi, a summer resort town, and in 1901 to Bakuriani, a winter-sports resort. Another line was completed in 1915 through the famous Georgian wine-producing region of Kakheti.[161]

Georgia's mountainous terrain presents difficult challenges for railway engineering, with over 1,500 kilometers of track through 45 tunnels and 1,716 bridges. Poti sits at only 6 meters above sea level, while the Marabda-Akhalkalaki section rises to over 2,200 meters above sea level. In total, five thousand major engineering structures, including breakwaters, flood banks, snow fences, and embankment protection, characterize the Georgian Railways. Today, Georgian Railways includes 115 freight stations, 8 locomotive depots, 4 wagon repair depots, 11 track branch depots, 7 power substation depots, 5 communication centers, locomotive and carriage works, railway police offices, 250 electric locomotives, about 200 diesel locomotives, and over 15,000 wagons. The existing track and rolling stock can handle 60 trains each day, corresponding to 40 million tons of freight each year.[162]

Construction of the Tbilisi-Kars rail line will help revive the economy of southern Georgia, and both Turkey and Georgia place a high priority on the opening of the rail line. Georgia has signed agreements with Uzbekistan, Turkmenistan, and Azerbaijan on freight transport and with Ukraine and Azerbaijan on transit passage. Georgian Railways is developing infrastructure for container and bulk freight in accordance with international requirements. Terminals at Batumi, Poti, and Tbilisi have been modernized for twenty- and forty-foot containers and the yards can handle up to ten thousand units per month. In general, the system specifications of the new modernized container terminals match European requirements due to EU technical assistance through TRACECA. In April 1999, a railway ferry opened between Poti and Batumi, Ilichevsk, and Varna.[163] The European Bank for Reconstruction and Development (EBRD) provided a $6 million grant to rehabilitate Georgian Railways' signaling and communications systems with fiber-optic technology. Some modernization of cargo-handling systems is still required. The container yards at Poti and Tbilisi are being expanded and rebuilt due to the growth in container traffic. Rehabilitation of signaling systems will be financed with $20 million from the EBRD and $5 million from the EU. The EU is also providing $1.2 million toward infrastructure on the Georgian-Azerbaijan railway, as well as $0.7 million to support rehabilitation of rolling stock and $1.5 million for establishing international timetables and tariffs. Georgian Railways is presently moving toward automatic gauge-changing axles.[164]

The Asian Development Bank–funded Masalli-Astara Highway project is part of the Southern Road Corridor Improvement project, focusing on construction, upgrade, and rehabilitation of 3,570 kilometers of 64 state roads and 5,928 kilometers of 58 secondary roads.[165] Azerbaijan joined the Asian Development Bank in 1999. This particular project includes not only construction of the fifty-nine-kilometer long, four-lane Masalli-Astara Highway, but also rehabilitation of local roads in the project area and installation of a vehicle-weighing station. Forty-five percent of total cumulative lending to Azerbaijan from the Asian Development Bank was allocated to the transport sector (about $810 million in November 2009), with 95 percent of that amount directed to road and highway improvement and development projects.[166] The Kuwait Fund and the Islamic Development Bank financed the upgrade of forty kilometers of the east-west road from Alyat to Gazi Mammad, and the EBRD considered a proposal to finance eighty kilometers of road from Gazi Mammad to Khurdamir. Road safety is still poorly developed.[167]

Georgian transit renovations have also benefited from donor financing. The EBRD funded a rehabilitation program at Tbilisi airport, which generally meets international standards on passenger and cargo carriage, but not in servicing. The internal air market of Georgia is highly monopolized, with 91 percent of the market share being dominated by a single company, as the entire market is contracting due to lack of competition. The American company Northrop Grumman has financed an upgrade of the air navigation system, which has increased transit flight capacities through Georgian airspace.[168] The upgrading of the Batumi airport with some Turkish financing has also improved cooperation between Georgia and Turkey. By constructing a complementary terminal in the town of Hopa, on the Turkish side of the border, Turkish citizens in the border regions are now able to travel to the rest of Turkey via the Batumi airport.[169]

Finally, in addition to physical transport, at least some progress has appeared in information technology and Internet access, in part due to the similarities between developing transnational rail linkages and developing transnational physical fiber-optic and other cable networks. The East-West Information Superhighway project will bridge two developed information networks in Western Europe and the Pacific and will pass through the Caucasus and Central Asia. The project aims to encourage development of the Internet, electronic information resources, and a knowledge-based economy in approximately twenty regions. In 2008, regional ministers signed the Baku Declaration on the Trans-Eurasia Information Superhighway.[170] Turkey, Azerbaijan, and Georgia lie on the Trans-Asia-Europe fiber-optic cable network. An underwater fiber-optic cable connects Poti in Georgia to Rize in Turkey, and additional cable follows the Georgian railway from Poti to the borders with Azerbaijan and Armenia. The European Union has contributed to the financing of the project.[171] Similarly, Georgia has the infrastructure to be able to export electricity to neighboring countries, and is connected via 220-, 330-, and 500-kilowatt electric transmission lines.[172] In 2005, Georgia adopted a new law on electronic communications that harmonizes the country with European and International Telecommunications Union standards, and the telecommunications sector is undergoing a rapid expansion in the country.[173] Similarly, a series of electronic commerce and privacy legislation in Azerbaijan shows great promise for economic development and harmonization with international standards.[174] Azerbaijan has the greatest Internet penetration rates among the South Caucasus countries, while Georgia has growth rates of over 20 percent per year.[175]

As with increasingly close cooperation in information-technology transfer and development, Azerbaijan in particular is leading the region's integration into the global telecommunications sector. Azerbaijan has launched two telecommunication satellites (Azerspace), with range covering a greater part of Eurasia and transmitting television, radio, Internet, international telephone, and other services.[176] Financing for Azerbaijan's first launch of a satellite, realized in 2013, came from the Export-Import Bank of the United States.[177] Only 20 percent of the satellite's telecommunication services will be reserved for Azerbaijan; the remainder will be sold on the international market.[178] Both Azerbaijan and Georgia have affirmed the SilkSat proposal of the East-West Space Science Center to create a comprehensive telecommunications satellite network for all of the Caucasus and Central Asia. Telecommunications infrastructure has deteriorated in the region since the fall of the Soviet Union, and difficult terrain hampers traditional physical telecom networks. SilkSat would augment the fiber-optic network developing in the region, and would boost trade among Caucasus and Central Asian states and boost foreign investment.[179] The governments of the region support a comprehensive regional satellite system.

Evaluation

Integrating transport policy among Azerbaijan, Georgia, and Turkey to the exclusion of their neighbors appears to be a slightly different and more decentralized process than energy integration. Whereas energy integration appeared to be a top-down phenomenon, facilitated by transnational actors and relatively insulated from popular pressures and other legal and regulatory frameworks, transport policy is a more complex phenomenon. Transport policy may have an impact on other legal and regulatory frameworks in a way that the insular nature of energy policy may not. Unlike energy interests in the three countries, for instance, transport policy seems more influenced by environmental issues, taxation and tariff structures, and visa and immigration policies. Similarly, communications and transactions among populations are considerably more important as a variable with regard to transport policy compared to energy policy. Passenger rail, for instance, has an effect on tourism, and the increasingly accessible nature of transnational highways and air passage reaches a broad segment of the population (or, in Deutsch's terms, at least politically relevant elites). The integration of telecommunications policy, even more so than rail, road, and air transit, has a strong impact on popular self-identity and perceptions of neighboring countries and peoples. At the same time, however, transnational extra-regional actors, such as international institutions, are not at all irrelevant. Many of the most prominent transport projects among Azerbaijan, Georgia, and Turkey have received financing and political support from international institutions. The European Union in particular views transport policy in the South Caucasus and Asia Minor as a critical element of its foreign policy toward the countries on the European periphery.

Ultimately, unlike energy integration, transport integration among Azerbaijan, Georgia, and Turkey requires a hybrid theory that draws elements of neofunctionalism, security communities theory, and neoliberal institutionalism. Unlike energy policy, transport integration is generally not isolated from other legal or regulatory processes. As a consequence, the entanglement of regulatory structures across borders may be expected, particularly as elements such as railroad gauge, air-traffic controls, environmental precautions, and road safety are relatively internationalized. Similarly, security communities theory is relevant because transport policy affects ordinary citizens as well as small and medium-sized business interests, and it shapes public perceptions and expectations of state capacity and relations with neighboring countries. As with energy policy, transnational transport and telecommunications policy in the South Caucasus and Asia Minor is facilitated to some extent by supranational actors such as the European Union, which perceives a transit corridor through the three countries as a central element of its wider Eurasia policy. Measuring the extent of integration among Azerbaijan, Georgia, and Turkey with respect to transport policy will have to account for all three variables: regulatory entanglement, popular communications and transactions, and facilitation by transnational extra-regional actors.

TRANSNATIONAL EXTRA-REGIONAL ACTORS

The European Union appeared soon after independence in the South Caucasus with the Technical Assistance to the Commonwealth of Independent States (TACIS) program, initiated in December 1990. As relations with the former Soviet republics moved higher on the European agenda in the succeeding years, the TACIS program spawned TRACECA (Transport Corridor Europe-Caucasus-Asia) and INOGATE (Interstate Oil and Gas Transport to Europe), the primary manifestations of the EU as a transnational extra-regional actor, or TERA, in the energy and transport spheres among Azerbaijan, Georgia, and Turkey. TRACECA, launched in 1993, aims to facilitate the region's access to world markets by developing a transport corridor as an alternative to a route though Russia. INOGATE, launched in 1995, aimed to create a favorable environment for private investment in the oil and gas sectors, including rehabilitation of existing infrastructure.[180]

The European Union's separate interactions with both Georgia and Turkey have contributed to de facto cooperation between the two countries on issues of trade, taxation, energy, and transport, through the two separate mechanisms of accession negotiations and the so-called European Neighborhood Policy.[181] With a number of foreign policy tools in its arsenal, including accession and partnership levels, as well as support for policy reform, the European Union has managed to bring about gradual social, political, and economic change to its immediate periphery. The EU's role as a constitutive agent is especially visible in the areas of trans-border relations, energy, and transport, due to EU-driven policy reforms in these areas. Similarities in these policy reforms facilitate interactions between the two countries as they eliminate technical issues that increase transaction costs. In addition, Mutlu writes, association with the EU creates a sense of belonging as to European practices on both sides of the border, creating a social capital that contributes to improved relations.[182] Although Turkey has a long history of cooperation with the EU, a formal agreement as to the EU membership process only began in 2005. Georgia's increasingly close EU relationship is even more recent: the country signed a Partnership and Cooperation Agreement in 1999, and the more recent European Neighborhood Policy and Partnership Instrument in 2006 and 2007, respectively, which aim to develop deeper economic integration. The EU is tasked with providing expertise and guidance to the Georgian government on judicial and legislative matters, including prosecution and organized crime and standardization of consumer goods. The Rose Revolution led to the "Europeanization" of Georgia's sociopolitical and economic landscape, and this rapid transformation has closed the gap between Turkey and Georgia on many issues.[183]

Although neither Georgia nor Turkey are full-fledged members of the EU, their commitment to important European projects, such as the BTC oil pipeline and the Southern Gas Corridor (often casually called "Nabucco"), has demonstrated that they "belong" to Europe in certain issue areas, as Europe represents higher standards and good governance to political elites in those countries. "Regardless of the form of connection, this sense of belonging, when combined with the other economic, social and political factors contributing to improved relations, can be seen as strong enough of a motivator to accelerate the emergence of a complex interdependence structure."[184]

After TRACECA's launch in Brussels in May 1993, the European Union–led technical assistance program has resulted in closer cooperation and dialogue among government authorities, and has led to agreements to keep transit fees at competitive levels and to simplify border-crossing formalities. The TRACECA corridor is the shortest and likely the fastest and cheapest transit route for large volumes of cargo from Central Asia to deepwater ports linked to world markets. TRACECA's technical assistance program has helped attract large investments from international financial institutions, including the EBRD's commitments for capital projects on ports, railways, and roads along the TRACECA route.[185] Likewise, the World Bank has made commitments for new capital projects on roads in Armenia and Georgia totaling over $40 million, and the Asian Development Bank has also committed substantial funds to road and railway improvements. The technical assistance has included setting up working groups to analyze the current system of border-crossing procedures; to determine unified rates of road-transit, port, and maritime fees; and to harmonize technical standards for road-vehicle characteristics and operations. In the first ten years, TRACECA financed thirty-nine technical assistance projects and fourteen investment projects for the rehabilitation of infrastructure, totaling about €110 million.[186] The integration and harmonization of the region's transport regulatory environment with European and international norms is an ongoing process.

The EU-supported TRACECA program is seeing high growth in container shipping. Mitsubishi is building a modern container berth at Poti to handle up to 300,000 containers per year. The most effective way of shipping freight between Tashkent in Uzbekistan to Poti in Georgia would be to run regular container services across the Caspian to Azerbaijani and Georgian rail networks. Poti is also a site of close linkages to railway ferries across the Black Sea to Ilichevsk, Ukraine, bypassing the Bosporus. TRACECA was instrumental in establishing ferry services between Poti and ports in Bulgaria and the Russian Federation. Poti also has container-shipping services to ports in Istanbul, Greece, Italy, and Western Europe.

The building of the BTC pipeline was an important goal of foreign donors in the region, and international financial institutions contributed handsomely to the project. Companies involved in the BTC pipeline project needed US intervention to secure support from local governments and funding from international financial institutions. With American encouragement, the International Finance Corporation (IFC), the private-sector arm of the World Bank, agreed to participate in the BTC project in November 2003.[187] The pipeline no doubt aligns with the broader regional goals of the international financial institutions and has led to significant regulatory reform at least in the energy sectors (and in pipeline security, taxation, and tariff structures) among the three countries. Despite its lack of a common energy-security policy, the European Union has always understood that Georgia would become a key transit country for European oil and gas imports, and included the country in the EU's 1997 Black Sea Strategy, which stressed, among other issues, the need to connect trans-European transportation networks with Caucasus infrastructure.[188] The European Union has sought to provide a collective legal framework for cooperation in oil and gas infrastructure sectors under INOGATE.[189]

The EU is by no means the only TERA operating in Azerbaijan, Georgia, and Turkey. The Organization for Security and Cooperation in Europe, of which Georgia has been a member since 1992, has participated in a long-term observer mission and mediation efforts in South Ossetia.[190] Since the mid-1990s, the OSCE gradually intensified its activities promoting liberal norms in Georgia through the OSCE Office for Democratic Institutions and Human Rights, which provided support in rewriting the Georgian constitution.[191] The OSCE, with support from the EU and other international organizations, assisted in preparing and conducting parliamentary and presidential elections and organized international monitoring efforts. Although none of the polls conducted during Shevardnadze's presidency met international standards, by 2003 the OSCE had established in Georgia one of the largest and longest-in-duration election-observer missions that ultimately revoked legitimacy from the election and from Shevardnadze's government. After the Rose Revolution, the OSCE turned its attention to strengthening democratic institutions in the realms of election capacity, legislative and local government reform, free media, and rule-of-law promotion.[192] Given the high interest of donors after the Rose Revolution in Georgia, the risk is high of overlap, failed expectations, or overstraining of political institutions. International donor agencies have only partially succeeded thus far in consolidating Georgian democracy.

TERAs have had a better track record in integrating the highly regulated transit and energy sectors than they have in integrating political institutions and promoting deep governance reforms. One reason may be that highly regulated sectors lend themselves to transnational standardization; an international organization may have more success, for instance, in improving customs and tariff procedures than in promoting judicial independence. Another reason may be that the energy and transit sectors have market incentives to internationalize, and the EU, OSCE, and other actors are simply promoting these market incentives. The European Commission's transit program for the South Caucasus has been especially successful. As export routes were being planned in the South Caucasus, the Commission began a new initiative: an alternative, temporary oil railway system between Azerbaijan and across TRACECA. Through a multilateral technical agreement, this system is designed to link European transport networks to Central Asia and the Caucasus and export extra Caspian oil that is not transported by pipeline.[193] As part of the TRACECA program, Azerbaijan Railways restructured and improved its control system in line with the TRACECA project in 1999, and constructed a new service depot funded by the European Union in order to upgrade three hundred kilometers of the Baku-Beiuk-Kjiasik rail line. In addition, the EBRD provided Georgia with a credit line of $20 million to modernize railway technology and the EU provided $6 million to lay an optical-fiber network.[194]

A goal of all eight Caucasus and Central Asian states is to expand the capacity for trade intra- and interregionally. TRACECA has worked with the EU and international lending institutions to obtain technical assistance for highway construction, port rebuilding, railway modernization, and other projects to shore up regional transport infrastructure.[195] The TRACECA project received financial support from the International Monetary Fund, the European Bank of Reconstruction and Development, and the International Bank of Reconstruction and Development at the World Bank. TRACECA has provided a forum for regional cooperation; for instance, five working groups have been held since the inception of the program, and a major conference held in Baku in 1998 called for ensuring access to the world market by road, rail, and commercial navigation. It also encouraged cooperation on traffic safety, environmental protection, legal transport regulation, and related goals. TRACECA has served as a bridge between the Caspian and the rest of the world around Russia, such as through a major rail link from Kazakhstan to China that can cut off several days compared to old Soviet routes.[196]

TERAs play a particularly important role in facilitating integration among Azerbaijan, Georgia, and Turkey to the exclusion of Russia, Armenia, and Iran. By funneling money to projects that align with the supranational agency's strategic goals, the agency can alter public perceptions and spur legal and regulatory reform. The European Union, for instance, has supported a series of projects that align with its broader strategic goals of creating a Southern Gas Corridor and transit route through Central Asia, the Caucasus, and Turkey. This has not led to political integration, as Turkey is not yet an EU member and Georgia is having difficulty even getting on track for membership. But public perceptions have no doubt changed as a result of the European Union's visible presence in Azerbaijan, Georgia, and Turkey. And perhaps just as importantly, the influx of donor money has led to increased pressure for administrative transparency, impartial legal oversight, and state-sponsored security for the projects.

CONCLUSION

This chapter has aimed to provide an introduction to integration among Azerbaijan, Georgia, and Turkey, with consideration of the three major theories of nation-state integration. Security communities theory, buoyed by the constructivist revolt against realism in international relations, holds that countries form bonds based on social perceptions of individual and collective self-identity. Communications and transactions among populations help to alter patterns of behavior and change expectations. Neofunctionalist theory identified a different mechanism for integration: the expansive nature of rational bureaucracy that would lead to regulatory and administrative intertwinement between countries. Although neofunctionalist theory has walked back from the original concept of spillover, in which regulatory integration in one sector would place pressure on other sectors, encouraging integration, neofunctionalism is still a deterministic or progressive theory in the sense that regulatory or administrative intertwinement creates its own momentum encouraging further integration. Neoliberal institutionalist theory, by contrast, holds that supranational actors will facilitate integration on their own terms and eventually alter state behavior and public perceptions as the integration deepens.

Among the countries of the South Caucasus, a heterogeneous region heavily shaped by a long period of Russian rule followed by a tumultuous independence, regional integration is a means for Georgia and Azerbaijan to overcome the downward economic spiral faced by both countries for much of the 1990s. Turkey is a willing partner, in part because its own energy needs make investment in Azerbaijan particularly attractive, and in part because its strategic location on the fringe of Europe gives it the advantage of being a Eurasian "bridge" of sorts. Integration, at least in the energy and transport sectors, is possible because of a division of labor among the three countries. Azerbaijan, as a producer and exporter of energy, has a particularly powerful voice in regard to regional energy policy, and its preferences favoring Georgia and Turkey over Armenia and Russia have sharply determined the course of energy integration in the region. Georgia, as a "transit" country with deepwater ports, a relatively well-developed rail and road infrastructure, and common borders with both Azerbaijan and Turkey, has great say over regional transport policy, including regulations on tariffs, freight, customs, and passenger mobility and safety. For its part, Turkey is a "donor" country, a financial sponsor for Georgia and Azerbaijan, and it has deep and important links to the European Union, international financial institutions, and the United States. Integration is possible among Azerbaijan, Georgia, and Turkey in part because each of the three countries possesses leverage over the other two that has made cooperation in policy easier and more successful. Each of the three countries will benefit from cooperating with the other two.

Political, security, and sociocultural integration have lagged far behind energy and transport integration in the region. This is not accidental. Energy and transport are both highly regulated sectors with involvement by multinational corporations and supranational institutions. The explanatory variable, illustrated in the following two chapters, is the role of transnational extra-regional actors in integrating the energy and transport sectors. As between the two of them, however, integration in the energy sector and in the transit sector have not proceeded in the same manner. Energy integration is highly facilitated from the top down and is removed from other realms of policy making and state regulation. Public perceptions play a comparatively smaller role than in transport policy. As a consequence, major energy decisions are often opaque if not secretive, and occur outside the formal political sphere. Transport integration, by contrast, is highly public, and it impacts other closely related fields such as immigration, customs, tariff and trade policy, and public-infrastructure finance such as selling government bonds. The differences between the two fields determine which integration theory is most appropriate for explaining why integration is occurring and the course it might take. Neoliberal institutionalism is particularly relevant for energy integration, because it best explains how a highly facilitated, top-down integrative process is helping to change state behavior with relatively little input from the general population. Transport integration is best explained by a hybrid theory, because it encompasses elements of bureaucratic and regulatory entanglement, popular perceptions as manifested in communications and transactions across borders, and by financing and leadership from supranational institutions, the three essential cores of the theories explained in the previous chapter. The following chapters will continue to review and refine these theories, complicating the original models while retaining the essential core.

NOTES

1. Soli Ozel (professor of international relations, Istanbul Bilgi University), interview with the author, April 17, 2007.

2. "Turkey's Caucasus Stability Initiative Driven by Own Interests," Azerbaijan Press Agency, September 15, 2009.

3. Yalçın Sarıkaya, "Georgian Foreign Policy after the August 2008 War," *Karadeniz Arastirmalari* 31 (2011): 11.

4. Angeliki Spatharou, "Geopolitics of Caspian Oil: The Role of Integration of the Caspian Region into World Economy in Maintaining Stability in the Caucasus," in *The Politics of Caspian Oil*, ed. Bulent Gokay (New York: Palgrave, 2001), 25, 33. Spatharou

also explains how, following the oil embargo, Armenia used images of food shortages and rolling blackouts, as well as a common affinity with Christian populations in the West, to shape public opinion in the United States. The efforts of Armenia, backed by a large diaspora, caused the United States to avoid recognizing independent Azerbaijan even as American oil companies were about to secure an 80 percent stake in the country's oil development. Ibid., 34.

5. "Russia Builds Second Military Base in Armenia," *Ayna*, May 3, 2008.

6. Avedis Bedros Hadjian, "Azerbaijan's Energy Policy and Its Implications for Russian Security," in Gokay, *Politics of Caspian Oil*, 139–40.

7. Ibid.

8. Ibid., 135.

9. Ibid., 136.

10. "Several Ethnic Azeri Activists Arrested in Iran," Turan News Agency, July 31, 2008.

11. Brenda Shaffer, "Iran's Internal Azerbaijani Challenge: Implications for Policy in the Caucasus," in *The Caspian Region*, vol. 1, *A Re-emerging Region*, ed. Moshe Gammer (New York: Routledge, 2004), 125.

12. Ibid., 127.

13. Gareth M. Winrow, *Turkey and the Caucasus: Domestic Interests and Security Concerns* (London: Royal Institute of International Affairs, 2000), 1.

14. Mithat Çelikpala, "From a Failed State to a Weak One? Georgia and Turkish-Georgian Relations," *Turkish Yearbook of International Relations* 36 (2005): 189.

15. Ibid., 190.

16. Ibid., 191.

17. Gulshan Pashayeva, "The Nagorno-Karabakh Conflict in the Aftermath of the Russia-Georgia War," *Turkish Policy Quarterly* 8, no. 4 (2009): 62.

18. "Azeri Think-Tank Group Looks into Projects Aimed at Stability in the South Caucasus," Turan News Agency, October 14, 2008.

19. Winrow, *Turkey and the Caucasus*, 5.

20. Tadeusz Swietochowski, *Russia and Azerbaijan: A Borderland in Transition* (New York: Columbia University Press, 1995), 218–19.

21. Damien Helly and Giorgi Gogia, "Georgian Security and the Role of the West," in *Statehood and Security: Georgia after the Rose Revolution*, ed. Bruno Coppieters and Robert Legvold (Cambridge, MA: MIT Press, 2005), 274.

22. Ministry of Foreign Affairs, Georgia, "Strategic Goals and Objectives," in *Foreign Policy Strategy 2006–2009* (Tbilisi: Ministry of Foreign Affairs, 2006), 7–17.

23. "Azeri Daily Criticizes Turkey's Handling of Georgian-Russian Conflict," *Zerkalo*, August 21, 2008.

24. Igor Torbakov, *The Georgia Crisis and Russia-Turkey Relations* (Washington, DC: Jamestown Foundation, 2008).

25. Çelikpala, "From a Failed State to a Weak One?," 188.

26. Ibid.

27. "Turkey-Azerbaijan Military Cooperation Deepens to the Tune of $200 Million," Azerbaijan Press Agency, May 13, 2010.

28. Svante E. Cornell, "Democratization Falters in Azerbaijan," *Journal of Democracy* 12, no. 2 (2001): 117–18; David Darchiashvilli, "The Arms Building and Security Problems in Georgia," NATO Fellowship Report (Brussels: NATO, 1997).

29. Turkey, or both Turkey and Russia, have membership in the Conference on Disarmament, the Nuclear Test Ban Treaty Organization, the Wassenaar arms-export organization, the Nuclear Suppliers Group, the Australia Group on chemical and biological weapons, and the Missile Technology Control Regime. Neither Azerbaijan nor Georgia have membership in any of these organizations.

30. "Armenian Expert Downplays Turkey's Role in Caucasus and Central Asia," Noyan Tapan News Agency, October 5, 2001.

31. Commission of the European Communities, *European Neighborhood Policy: Country Report, Georgia,* Commission Staff Working Paper (Brussels: Commission of the European Communities, 2005), 12.

32. Svante E. Cornell, "Turkey and the Conflict in Nagorno Karabakh: A Delicate Balance," *Middle Eastern Studies* 34, no. 1 (1998): 51.

33. "Turkey Not to Resume Ties with Armenia until Karabakh 'Occupation' Ends," Anatolian Agency, June 10, 2012.

34. Ibid.

35. Spatharou, "Geopolitics of Caspian Oil," 29.

36. Ibid., 31–32.

37. Mehmet Tütüncü, "Turkish Foreign Policy in the Caucasus," in *Contrasts and Solutions in the Caucasus,* ed. Ole Høiris and Sefa Martin Yürükel (Oakville, CT: Aarhus University Press, 1998), 480.

38. Yuri N. Zinin and Alexei V. Maleshenko, "Azerbaijan," in *Central Asia and the Caucasus after the Soviet Union: Domestic and International Dynamics,* ed. Mohiaddin Mesbahi (Gainesville: University Press of Florida, 1994), 114.

39. Ibid.

40. Thomas De Waal, *The Caucasus: An Introduction* (Oxford: Oxford University Press, 2010), 182.

41. Kamer Kasim, "The Nagorno-Karabakh Conflict, Caspian Oil and Regional Powers," in Gokay, *Politics of Caspian Oil,* 187–88.

42. Winrow, *Turkey and the Caucasus,* 8.

43. Ibid., 10.

44. Kasim, "Nagorno-Karabakh Conflict," 190–91.

45. Winrow, *Turkey and the Caucasus,* 10.

46. Idris Bal, *Turkey's Relations with the West and the Turkic Republics: The Rise and Fall of the "Turkic Model"* (Burlington, VT: Ashgate, 2000), 205–6.

47. Ibid.

48. Nazrin Mehdiyeva, *Power Games in the Caucasus: Azerbaijan's Foreign and Energy Policy Towards the West, Russia and the Middle East* (New York: Taurus, 2011), 185–86.

49. David I. Hoffman, "Azerbaijan: The Politicization of Oil," in *Energy and Conflict in Central Asia and the Caucasus,* ed. Robert Ebel and Rajan Menon (New York: Rowman & Littlefield, 2000), 65.

50. Hoffman, "Azerbaijan," 67.

51. Darrell Slider, "Democratization in Georgia," in *Conflict, Cleavage, and Change in Central Asia and the Caucasus,* ed. Karen Dawisha and Bruce Parrott (New York: Cambridge University Press, 1997), 185, 190.

52. Thomas De Waal, "Georgia and Its Distant Neighbors," in Coppieters and Legvold, *Statehood and Security,* 314.

53. Gareth M. Winrow, *Turkey and Caspian Energy* (Abu Dhabi: Emirates Center for Strategic Studies and Research, 1999), 9.

54. Winrow, *Turkey and the Caucasus,* 15.

55. Ibid., 16.

56. "Address by H. E. Ahmet Davutoğlu, Minister of Foreign Affairs of Republic of Turkey at the OSCE Ministerial Council Meeting, 1–2 December 2009, Athens," *Turkish Government News,* July 6, 2010, http://infoweb.newsbank.com/resources/doc/nb/news/13125419D157C608?p=AWNB.

57. Fiona Hill, "Seismic Shifts in Eurasia: The Changing Relationship between Turkey and Russia and Its Implications for the South Caucasus," *Journal of Southeast European and Black Sea Studies* 3, no. 3 (2003): 66, 70.

58. De Waal, "Georgia and Its Distant Neighbors," 324–25.

59. Ibid., 314. De Waal notes that Azerbaijan and Russia have experienced a thaw in their relations at a time when Georgia-Russia tension was increasing. For more on Turkey's burgeoning economic relations with Russia, see Winrow, *Turkey and the Caucasus,* 1.

60. Georgia: 72 percent viewed themselves as Georgian; 17 percent as Georgian and European; and 2 percent as European. The Caucasus Research Resource Centers, "Caucasus Barometer 2008," http://www.crrc.ge/oda. When the question is framed differently, however, the results are striking. In 2011, when asked whether they agreed with the sentiment "I am Georgian, therefore I am European," 59 percent agreed, while only 31 percent did not agree. Ibid., "EU Survey 2011."

61. Ibid.

62. "Turkey's Caucasian Adventure," *Mideast Mirror*, February 2, 2000.

63. For instance, survey data show that public support for supranational organizations is skeletal at best. In Azerbaijan, trust toward the European Union was highly mixed (15 percent strongly distrust, 17 percent distrust, 25 percent neutral, 17 percent trust, and 5 percent strongly trust). Membership in NATO polled considerably better. In Georgia, trust toward the European Union was considerably higher (4 percent strongly distrust, 6 percent distrust, 25 percent neutral, 31 percent trust, and 13 percent strongly trust). This data coincides with the above theory that Georgia possesses a stronger Western outlook than Azerbaijan does. The Caucasus Research Resource Centers, "Caucasus Barometer 2010 (Georgia and Azerbaijan)," http://www.crrc.ge/oda.

64. De Waal, *The Caucasus*, 7–8.

65. Revaz Gachechiladze, "Geopolitics in the South Caucasus: Local and External Players," *Geopolitics* 7, no. 1 (2002): 114.

66. Ibid., 116.

67. Ibid., 118.

68. De Waal, *The Caucasus*, 16.

69. Elaheh Koolaee and Mohammed Hossein Hafezian, "The Islamic Republic of Iran and the South Caucasus Republics," *Iranian Studies* 43, no. 3 (2010): 398.

70. Ibid.

71. "Azeri Surprise at Iran's Protest over Caspian Oil Exploration Projects," Turan News Agency, July 23, 2001.

72. Koolaee and Hafezian, "Iran and the South Caucasus Republics," 399–401.

73. Brenda Shaffer, *Borders and Brethren: Iran and the Challenge of Azerbaijani Identity* (Cambridge, MA: MIT Press, 2002), 209.

74. Audrey L. Altstadt, "Azerbaijan's Struggle toward Democracy," in Dawisha and Parrott, *Conflict, Cleavage, and Change*, 145.

75. Shaffer, *Borders and Brethren*, 209.

76. Zinin and Maleshenko, "Azerbaijan," 109–11.

77. Shaffer, *Borders and Brethren*, 211.

78. Ibid., 212.

79. Emmanuel Karagiannis, *Energy and Security in the Caucasus* (New York: Routledge, 2002), 144.

80. Ibid., 145–46.

81. Altstadt, "Azerbaijan's Struggle toward Democracy," 143.

82. Slider, "Democratization in Georgia," 193–94.

83. Martha Brill Olcott, "Regional Cooperation in Central Asia and the South Caucasus," in Ebel and Menon, *Energy and Conflict*, 134.

84. Iveta Svota, Mark S. Johnson, and Stephen Heyneman, "Education and the Crisis of Social Cohesion in Azerbaijan and Central Asia," *Comparative Education Review* 51, no. 2 (2007): 159–80.

85. See S. Frederick Starr and Svante E. Cornell, eds., *The Baku-Tbilisi-Ceyhan Pipeline: Oil Window to the West* (Washington, DC: Central Asia–Caucasus Institute, 2005)

86. Tanja A. Börzel and Yasemin Pamuk, "Pathologies of Europeanisation: Fighting Corruption in the Southern Caucasus," *West European Politics* 35, no. 1 (2012): 79–97.

87. This interesting observation is from Tamar Shioshvili, "Prospects of Entering the Global World of American Studies in Georgia," paper presented at "New Trends in Higher Education," Fourth International Silk Road Symposium, International Black

Sea University, Tbilisi, Georgia, May 4–6, 2007, available at http://silkroad.ibsu.edu.ge/previous/fourth/7_10.pdf.

88. Economic statistics derive from the statistics database of the World Trade Organization, with numbers from October 2011, available at http://stat.wto.org.

89. "Turkey's Energy Strategy," *Turkish Government News*, July 6, 2010.

90. Samuel Lussac, "The Baku-Tbilisi-Kars Railroad and Its Geopolitical Implications for the South Caucasus," *Caucasian Review of International Affairs* 2, no. 4 (2008): 214.

91. Yigal Schleifer, "Turkey: Caucasus Crisis Leaves Ankara Torn between U.S. and Russia," *EurasiaNet*, November 9, 2008.

92. Matthias Luecke and Natalia Trofimenko, "Whither Oil Money? Redistribution of Oil Revenue in Azerbaijan," in *The Economics and Politics of Oil in the Caspian Basin: The Redistribution of Oil Revenues in Azerbaijan and Central Asia*, ed. Boris Najman, Richard Pomfret, and Gael Raballand (New York: Routledge, 2008), 134.

93. De Waal, "Georgia and Its Distant Neighbors," 314–15.

94. Ibid.

95. Ibid., 317.

96. Fiona Hill, "Seismic Shifts in Eurasia," 66.

97. Fiona Hill and Omer Taspinar, "Turkey and Russia: Axis of the Excluded?," *Survival* 48, no. 1 (2006): 84, 87.

98. De Waal, *The Caucasus*, 189–90.

99. Ibid., 195.

100. De Waal, "Georgia and Its Distant Neighbors," 316.

101. Ibid., 317.

102. Lincoln A. Mitchell, "Democracy in Georgia since the Rose Revolution," *Orbis* 50, no. 4 (2006): 670, 672.

103. David S. Siroky and David Aprasidze, "Guns, Roses, and Democratization: Huntington's Secret Admirer in the Caucasus," *Democratization* 18, no. 6 (2011): 9–10, 12.

104. Can E. Mutlu, "A De Facto Cooperation? The Increasing Role of the European Union in Improved Relations between Georgia and Turkey," *Comparative European Politics* 9, nos. 4–5 (2011): 556.

105. "Russia's Top Law-Enforcement Officials Discuss Caucasus Drug Trafficking," Interfax, February 25, 2010.

106. De Waal, "Georgia and Its Distant Neighbors," 316.

107. Olcott, "Regional Cooperation in Central Asia and the South Caucasus," 130–31.

108. Vladimir Papava, "Georgia's Economic Role in the South Caucasus," *Problems of Economic Transition* 48, no. 4 (2005): 89.

109. These statistics, as those above, are derived from the World Trade Organization's statistics database at http://stat.wto.org.

110. "Conference in Azeri Capital Looks into Ways of Accession to WTO," *Zerkalo*, October 28, 2010.

111. This information is derived from the European Commission, "South Caucasus," http://ec.europa.eu/trade/creating-opportunities/bilateral-relations/regions/south-caucasus.

112. "Russian TV Portrays GUAM Summit as US-Funded Bid to Weaken CIS," *Vesti*, May 23, 2006.

113. De Waal, *The Caucasus*, 167.

114. Ibid.

115. Ibid., 169.

116. Luecke and Trofimenko, "Whither Oil Money?," 134.

117. Spatharou, "Geopolitics of Caspian Oil," 28–29.

118. Hadjian, "Azerbaijan's Energy Policy," 131–32.

119. Ibid., 132–33.

120. Hoffman, "Azerbaijan," 58.

121. Ibid., 59.

122. Ibid., 60–61.
123. Ibid., 63.
124. Spatharou, "Geopolitics of Caspian Oil," 31–32.
125. Winrow, *Turkey and the Caucasus*, 27–28.
126. Gachechiladze, "Geopolitics in the South Caucasus," 127.
127. De Waal, "Georgia and Its Distant Neighbors," 322–23.
128. Gachechiladze, "Geopolitics in the South Caucasus," 127.
129. De Waal, *The Caucasus*, 181.
130. Ibid.
131. Charles van der Leeuw, *Oil and Gas in the Caucasus and Caspian: A History* (New York: St. Martin's, 2000), 134.
132. Ibid., 139.
133. Hooshang Amirahmadi, "Challenges of the Caspian Region," in *The Caspian Region at a Crossroad: Challenges of a New Frontier of Energy and Development*, ed. Hooshang Amirahmadi (New York: St. Martin's, 2000), 16.
134. Van der Leeuw, *Oil and Gas in the Caucasus and Caspian*, 130.
135. Winrow, *Turkey and Caspian Energy*, 25.
136. De Waal, *The Caucasus*, 184–85.
137. Ibid., 185.
138. Ibid., 185–86.
139. Ibid., 186–87.
140. Ibid., 182–83.
141. Luecke and Trofimenko, "Whither Oil Money?," 137.
142. Ibid., 151–52.
143. Pirouz Mojtahed-Zadeh, "The Geo-politics of the Caspian Region," in Amirahmadi, *Caspian Region at a Crossroad*, 180–81.
144. Gachechiladze, "Geopolitics in the South Caucasus," 125.
145. Bülent Aras and George Foster, "Turkey: Looking for Light at the End of the Caspian Pipeline," in *Oil and Geopolitics in the Caspian Sea Region*, ed. Michael Croissant and Bülent Aras (Westport CT: Praeger, 1999), 234–35.
146. Aras and Foster, "Turkey," 236.
147. Van der Leeuw, *Oil and Gas in the Caucasus and Caspian*, 128–29.
148. Lussac, "Baku-Tbilisi-Kars Railroad," 212.
149. Ibid., 213.
150. The completion of the project was delayed once to 2013 and again to late 2015 or 2016.
151. Ibid.
152. Gachechiladze, "Geopolitics in the South Caucasus," 113–14.
153. Lussac, "Baku-Tbilisi-Kars Railroad," 215.
154. Ibid., 216.
155. Winrow, *Turkey and the Caucasus*, 16.
156. Rovshan Ismayilov, "Azerbaijan, Georgia, and Turkey: Building a Transportation Triumvirate?," *EurasiaNet*, June 2, 2007.
157. Lussac, "Baku-Tbilisi-Kars Railroad," 217.
158. "Kars-Tbilisi-Baku Railway Construction to Begin with Ceremony in Tbilisi 20 Sep," Anatolia News Agency, August 13, 2007,
159. Lussac, "Baku-Tbilisi-Kars Railroad," 217.
160. Mamuka Tsereteli, *The Impact of the Russia-Georgia War on the South Caucasus Transportation Corridor* (Washington, DC: Jamestown Foundation, 2009), 9.
161. Teimuraz Gorshkov and George Bagaturia, "Developments at Georgian Railway," *Japan Railway and Transport Review* 24 (2000): 42–43.
162. Ibid., 43–44.
163. Ibid., 46.
164. Ibid., 47.

165. Irshad Abbasov and Elnur Ahmadov, "Mainstreaming Climate Change Impacts of Construction of 59-km long Masalli-Astara Highway Project," *Bankwatch* 8, no. 1 (2010): 12–15.
166. Ibid.
167. Azeravtoyol, "Azerbaijan Highway Project: Project Initiation Document," March 27, 2001.
168. Archil Gegeshidze, "Georgia's Function as a Transit Country and Sustainable Development," presentation at the World Summit on Sustainable Development, Johannesburg, August 26–September 4, 2002.
169. Mutlu, "A De Facto Cooperation?," 554.
170. Rufat H. Gulmammadov, "The National ICT Strategy and Key Indicators on Information Society of Azerbaijan Republic," paper presented at the e-GovShare Conference, Antalya, Turkey, December 8–11, 2009.
171. Gegeshidze, "Georgia's Function as a Transit Country."
172. Ibid.
173. Mariam Sulaberidze, "Global Information Society, Digital Divide, and Digital Opportunities of the South Caucasus Region," in *First Regional Students' Scientific Conference: Development Perspectives of South Caucasus Region* (Tbilisi: Caucasus University, 2008), 88.
174. Stephen E. Blythe, "Azerbaijan's E-Commerce Statutes: Contributing to Economic Growth and Globalization in the Caucasus Region," *Columbia Journal of East European Law* 1, no. 1 (2007): 44.
175. Alexy Sidorenko and Arzu Geybullayeva, "The Internet in Azerbaijan," *Caucasus Analytical Digest* 15 (2010): 4, and Alexy Sidorenko, "Internet, Society, and Democracy in Georgia," *Caucasus Analytical Digest* 15 (2010): 7.
176. Gulmammadov, "National ICT Strategy."
177. "US Ex-Im Approves Loan for Construction of Azerbaijani Satellite Azerspace," *Trend*, April 29, 2011.
178. Jeffrey Hill, "Azercosmos: Work Underway on New Azerspace Satellite," *Satellite Today*, April 10, 2012.
179. Lee Berger, "Proposed Legal Structure for the SilkSat Satellite Consortium: A Regional Intergovernmental Organization to Improve Telecommunications Infrastructure in Central Asia and the Trans-Caucasus Region," *Law and Policy in International Business* 33, no. 2 (2001): 100ff.
180. Yelda Demirag, "EU Policy towards South Caucasus and Turkey," *Perceptions* 9, no. 4 (2004): 92.
181. Mutlu, "A De Facto Cooperation?," 544.
182. Ibid., 546–48.
183. Ibid., 548–49.
184. Ibid., 549–50.
185. Bernard Touboul, "The Central Asia Harmonisation of Border Crossing Procedures in the Perspective of TRACECA," paper presented at the Regional Seminar on Trade Facilitation and Customs Modernization, Issyk-Kul, Kyrgyzstan, August 4–8, 2003.
186. Ibid.
187. Helly and Gogia, "Georgian Security and the Role of the West," 279–280.
188. Ibid.
189. Ibid., 280–81.
190. "Turkey Sends Two Military Observers to South Ossetia as Part of OSCE Mission," Anatolian Agency, September 4, 2008.
191. Pamela Jawad, "Conflict Resolution though Democracy Promotion? The Role of the OSCE in Georgia," *Democratization* 15, no. 3 (2008): 618.
192. Ibid., 620.
193. Helly and Gogia, "Georgian Security and the Role of the West," 280–81.
194. Teimuraz Gorshkov and George Bagaturia, "TRACECA: Restoration of Silk Route," *Japan Railway and Transport Review* 28 (2001): 54.

195. Olcott, "Regional Cooperation in Central Asia and the South Caucasus," 129.
196. Ibid.

THREE

Transnational Extra-Regional Actors: The World Bank Group

The role of international financial institutions in promoting development of the energy and transport sectors of Azerbaijan, Georgia, and Turkey presents an illustrative example of transnational extra-regional actors as the key variable in the integration of the sectors in the three countries. Although a number of financial institutions such as the Asian Development Bank (ADB), the International Monetary Fund (IMF), and the European Bank for Reconstruction and Development (EBRD) have key roles to play in the Caucasus and Caspian region, this chapter has chosen to primarily focus on the regional activities and attendant integrative effects of the World Bank Group. The following three sections provide an overview of the major development projects directly implemented or otherwise financed by the World Bank Group in Azerbaijan, Georgia, and Turkey, with a particular focus on the development of the oil and gas sectors. The final sections of the chapter discuss two particular characteristics of the Group—its unparalleled ability to coordinate with other extra-regional actors, such as donor and financial institutions and its promulgation of various ranking mechanisms—that provide harmonized benchmarks for the energy and transport sectors in Azerbaijan, Georgia, and Turkey. Because governments that accept funds from the World Bank must comply with a complete package of reforms, and cannot choose "à la carte" from a menu of reform options, the World Bank's activities promote harmonization that allows for regional integration in these sectors.[1]

OVERVIEW

The World Bank Group and its International Finance Corporation (IFC), the Bank Group's private-sector-investment arm, produce regular Country Partnership Strategy (CPS) reports and periodic updates on Azerbaijan, Georgia, and Turkey that outline completed and anticipated Bank projects and the disbursements allocated for their implementation. A significant portion of each of these CPS reports concerns the Bank's considerable involvement in the transport and energy infrastructure sectors of Azerbaijan, Georgia, and Turkey and the critical linkages between them.

The Bank's early support for two major pipeline projects lends insight into the value-added the organization provides on multiple levels to advance integration between the three countries in their energy sectors. BTC consortium and international financial institution officials concur that the BTC and Baku-Supsa pipelines, two critical energy transit projects in the region, were made possible by the early funding and support provided by the World Bank Group.[2] The IFC provided a $200 million loan for the construction of the Baku-Supsa pipeline and a $250 million overall investment in the development of the BTC pipeline, which was completed in July 2006 and required over $2.6 billion in public- and private-sector investments. In Azerbaijan alone, the IFC provided loans of over $92 million to help realize the BTC project and its attendant infrastructure.

Oversight by the World Bank ensured that the pipelines were built and are operated according to harmonized and integrated standards and environmental guidelines, and the Group's presence was critical in attracting commercial financiers to provide the follow-on funds for the pipelines' completion. The World Bank Group also assumed the role of international arbiter to ensure that the BTC consortium's production-sharing agreements with Azerbaijan, Georgia, and Turkey were carried out in an equitable manner. The Group's commitment to the BTC and Baku-Supsa projects also helped mitigate political risk factors that otherwise would have contributed to private-sector liabilities and consequently hindered the financial commitment of major international extraction companies. Above all, the legitimacy and supervision provided by the World Bank were critical to securing public recognition and support from the US and EU governments and international NGOs for the project. The above cases provide just a brief glimpse of the ways in which Bank (or transnational extra-regional actor) support for key energy infrastructure projects has helped to physically integrate the energy sectors of Azerbaijan, Georgia, and Turkey.

As discussed earlier, the World Bank Group's major stake and financial involvement in the development of the energy and transport sectors of Azerbaijan, Georgia, and Turkey has been, and continues to be, contingent upon the implementation of a number of good-governance and integrative measures in the three countries. In order to secure financing for the development of lucrative energy-related infrastructure projects, the governments of Azerbaijan, Georgia, and Turkey must participate in the same World Bank Group programs that promote good governance, rule of law, and financial transparency. This harmonization allows for integration. The potential social and environmental impact of energy-related projects is extensively reviewed by the Group as a precondition for receiving financing, and the Bank itself is regularly reviewed by independent authorities to ensure that it mitigates the downside risks of extractive-industry projects on local communities and aligns with good-governance practices and the overall poverty-reducing mission of the organization.[3] The presence of—and partnership with—the Group also helps ensure that domestic and foreign private-sector companies operating out of the region adhere to the same standards, respect environmental guidelines, and provide secondary benefits to the local community. As one senior World Bank official put it, the Group's programs in any one country represent a package that is "more than the sum of its parts." As the official explained, providing financing "opens up a lot of other areas that we [the World Bank Group] need to be involved in if we are to be assured that our money is well-used."[4]

AZERBAIJAN

Energy Sector Development and Local Community Outreach

Since the late 1990s, the World Bank Group has been intensively involved in the energy and transport-related sectors in Azerbaijan and has served as a catalyst to "crowd in" private investment. With respect to the BTC project, the Group worked closely with the Azerbaijani government on an economic-reform program designed to ensure an adequate level of transparency and accountability necessary to attract the international financing for such a mammoth undertaking. In addition to the completion of the BTC pipeline, the Group's ongoing Railway Transport and Trade Facilitation Project works to improve the efficiency and modernization of the national railway company, Azerbaijan Dovlet Demir Yolu. Railways are an important component of the infrastructural connections among Azerbaijan, Georgia, and Turkey, particularly as an ever-increasing amount of oil is transported across rail links in the country.[5] This project complements the Group's parallel efforts to improve railways in Georgia and upgrade integral energy transit routes from Baku to Georgia's Black Sea ports.

Although complex, high-risk infrastructure development initiatives like the BTC pipeline make up a relatively small fraction of the IFC's overall project portfolio, such projects tend to receive the most outside scrutiny from concerned parties and civil-society organizations whose aim is to ensure an adequate level of social and environmental supervision. Aside from providing funding and facilitating an enabling investment climate, the Group has convened forums for affected local communities and stakeholder organizations that would like to voice concerns and receive further information about the project. In many ways, the BTC pipeline became a test case for the World Bank's ability to achieve sustainable development outcomes through effective mobilization of diverse stakeholder interests. Prior to final consideration of the project by the IFC's board, the Group embarked on an extensive Environmental and Social Impact Assessment report that was among the most comprehensive and voluminous the Bank had ever produced—spanning forty-six volumes and eleven thousand pages—and included a 120-day public comment and disclosure period. The IFC worked with the BTC consortium to ensure that there was no physical displacement of households along the pipeline route, and to provide guidance on temporary land resettlement and compensation during the project's multiyear construction.

With regard to the longer-term benefits to be harvested from the completion of the pipeline, the IFC has worked with the BTC consortium and partner NGOs to develop a Community Investment Programme (CIP) for Azerbaijan, Georgia, and Turkey. The CIP is intended to help local communities reap the secondary socioeconomic benefits from the project and develop successful "linkage projects" related to energy efficiency, community healthcare, microfinance, and local infrastructure support.[6] The IFC also collaborates with private-sector partners like BP to magnify the positive impact of the construction of the BTC pipeline on local businesses and suppliers by providing grants to the Micro Finance Bank of Azerbaijan to expand its branch network in communities where BP conducts economic development operations. As one IFC official has noted, such an initiative is a "good example of how IFC can work to increase the value-added that large projects in the oil and gas sector can bring to host countries, their businesses, and local communities."[7]

State Oil Fund and Resource-Revenue Transparency

In order to optimize Azerbaijan's management of its resource wealth, the Group advised the Azerbaijani government on the creation of the State Oil Fund of Azerbaijan (SOFAZ).[8] SOFAZ was established in 1999 by presidential decree and incorporates the best management practices of other natural resource funds such as Norway's Government Petroleum Fund and Kuwait's Reserve Fund for Future Generations.[9] The oil fund is used to save energy-related windfalls for future generations and serve as a sterilization mechanism to prevent the onset of Dutch disease (for a more complete discussion of Dutch disease refer to the following section).

Annual audits and quarterly financial reports are issued regularly and contribute to transparency. Like many other sovereign funds, SOFAZ maintains a conservative investment strategy, investing primarily in high-quality, low-risk securities outside Azerbaijan.[10] A portion of the country's total energy revenues flows to SOFAZ, with the rest flowing directly to the government's budget.[11]

To promote broad-based and internationally recognized practices of disclosure and transparency in the management of Azerbaijan's energy revenues, the World Bank Group facilitated the country's participation in the Extractive Industry Transparency Initiative (EITI). EITI is a coalition of governments, investors, and civil-society groups that endorse improved governance in resource-rich countries by calling for full publication and independent verification of company payments and government revenue streams from natural resources. The initiative was launched at the behest of former UK Prime Minister Tony Blair at the 2002 World Summit on Sustainable Development and was piloted by the UK's then-development agency, the Department for International Development (DFID). The government of Azerbaijan signed a Memorandum of Understanding with a coalition of NGOs and major oil and gas companies in Baku in 2004, a first step toward Azerbaijan's becoming an EITI-compliant country. The National Committee on EITI is chaired by SOFAZ and oversees the implementation of EITI principles. Over the course of the past five years, ninety NGOs have been able to monitor energy revenues in Azerbaijan and develop a dialogue with the government about how such revenues are disclosed. Since 2007, the State Oil Company of Azerbaijan (SOCAR) has been subject to audit to International Financial Reporting Standards as part of the EITI criteria. To facilitate Azerbaijan's preparation for this auditing process, the Group has cooperated with the European Union's Technical Aid to the Commonwealth of Independent States (TACIS) program to implement financial accounting and reporting reform. In 2008, Ernst and Young published SOCAR's first internationally accepted audit report. These efforts represent an important step forward in SOCAR's gradual transformation from an opaque Soviet-era company into a more transparent modern state-owned oil and gas company along the lines of Norway's StatoilHydro and Brazil's Petrobras.[12] Both the United Nations and the Open Society Institute have commended Azerbaijan in recent years for its leadership role within the EITI and its achievement of being recognized as the first EITI-compliant country.[13]

Non–Oil Sector Development

The World Bank has noted the recent energy bonanza remains particularly problematic for Azerbaijan. The country's oil reserves are known to be more limited than is the case for many other oil exporters, and therefore the resource boom is likely to be more short-lived. In a 2005 report, the World Bank predicted that Azerbaijan's oil production could peak as early as 2010 at an estimated seventy-one million tons.[14] Furthermore, most proven and potential oil resources in Azerbaijan are located offshore, which tends to exacerbate the difficulty of their extraction.[15] The global financial crisis, which began in 2008, compounded Azerbaijan's vulnerability in this respect, as commodity prices plummeted with a slowdown in global demand.

In light of these realities and the looming specter of Dutch disease, a significant portion of the Group's activities is geared toward developing Azerbaijan's non-oil sector. Dutch disease, also known as the "resource curse," is a classic economic problem that afflicts resource-rich nations, wherein the national currency severely appreciates as a result of the resource wealth and results in attenuation of the manufacturing and non-resource-related industries of a country. To ameliorate these effects, SOFAZ serves as the main instrument of sterilization for the influx of foreign-exchange flows resulting from the oil boom that, if left unmanaged, could precipitate rapid appreciation and overheating of the economy.[16] Moreover, the oil revenues accumulated in SOFAZ represent an important source of financing for development of the non-oil economy.

In addition to the role of SOFAZ, the IFC has partnered with the private sector to spur foreign direct investment in non-oil-related industries such as financial services, utilities, and agribusiness. Development of Azerbaijan's non-oil sector is integral to achieving greater poverty reduction and job growth in Azerbaijan, since expanded petroleum output tends to have very little direct impact on employment. In 2001, the energy sector accounted for 30 percent of GDP but only 1 percent of employment, a discrepancy which has only been further exacerbated in recent years with new petroleum exports coming online and when oil prices rise.[17] The IFC's strategy for the non-oil sectors in Azerbaijan focuses on promoting competition in the banking sector, providing technical assistance to private local banks, improving access to credit, and creating the right enabling environment for private provision of public services.[18] The IFC works to achieve these objectives through specific initiatives like the Supplier Finance Facility (SFF) that provides capital to small and medium-sized enterprises.[19] The SFF is implemented through the Micro Finance Bank of Azerbaijan—the first of its kind in the country and established with the help of the IFC—and in June 2006 provided its first loan to Azmetco, a local BP supplier that specializes in metering services.[20]

Governance and Rule of Law

Corruption and government inefficiency remain one of Azerbaijan's most significant problems and continue to hinder efforts to attract greater international investment. Through the efforts of its International Development Association (IDA) and IFC arms, the World Bank Group has made improving governance and combating corruption one of its primary areas of focus in Azerbaijan. Despite some marginal improvements in past years, Azerbaijan has consistently ranked low on Transparency International's Corruption Perceptions Index (158th out of 180 countries surveyed in 2008).[21] In 2005, the Group's board approved a Poverty Reduction Support Credit to expand anticorruption efforts and improve the accountability and transparency of public finances.[22] The IFC's Corporate Governance Project offers ongoing technical assistance to banks and joint-stock companies in Azerbaijan that are interested in improving their corporate governance practices. The Bank and the IMF have also teamed up to assist the Azerbaijan National Bank with the drafting and passage of an anti-money-laundering law consistent with Financial Action Task Force guidelines and the creation of a financial intelligence unit to monitor illicit money flows. [23] The World Bank's E-Government and Public Sector Reform program supports the "modernization of public sector administrative systems," while the Judiciary Reform Project focuses on "strengthening the management capacity of judicial institutions, upgrading court facilities by building and rehabilitating courthouses and providing technology and equipment, strengthening the professionalism of judges and judicial sector staff by providing training and upgrading training facilities, and improving citizen access to information."[24]

GEORGIA

Energy and Transport Sectors

Since Georgia's entry as a member of the World Bank in 1992, the Group has bolstered the country's role as a transit economy and in recent years supported the reform-oriented policies of the post–Rose Revolution government.[25] Since the early nineties, Georgia has borrowed more than $1 billion from the World Bank's International Development Agency and remains one of the largest borrowers in the Europe and Central Asia region.[26] Although the Group's assistance to Georgia with its decades-long transformation from a Soviet command economy to a market-led one has been multifaceted, the results have been particularly striking in integrating its energy sector with its neighbors Azerbaijan and Turkey. In this and in other crucial industries, the World Bank acts as a catalyst to attract the necessary donor and private financing for a burgeoning transit corridor. In 1997, the Group agreed to fund an Oil Institution Building Project that supported the newly created Georgian International Oil Company in its overall objective of maximizing the potential benefits from accelerating oil and gas development in the Black Sea and Caspian region. The Bank credit financed a feasibility study for the Baku-Supsa export pipeline, as well as provided a set of trained international financial and legal advisers to assist with the pipeline negotiations.

In 1998, the IFC's board approved a $200 million loan for the Early Oil Project, a commitment to aid with the construction of Black Sea terminals and the Baku-Supsa pipeline in Azerbaijan and Georgia. The project represented the first major oil development in the region since the collapse of the Soviet Union, and was part of a larger phased project that integrated staged development of the Azeri-Chirag-Guneshli field operated by BP for a larger consortium of oil companies that held the licensing rights to exploration. The $200 million IFC financing went to five major foreign oil firms in the consortium "to mobilize necessary funds, reduce their exposure to a project with significant political risk, and lay the groundwork for future financing for related projects in the region."[27] In terms of the transit fees collected by the Georgian government from these and other projects like the BTC pipeline, the IMF estimates such revenues could top $50 million annually by 2011, a significant figure for an economy the size of Georgia's.[28] The early support of the World Bank helped underpin other legal reforms, supported by the 1999 Energy Sector Adjustment Credit, that encouraged private-sector investment in the Georgian oil and gas industry and facilitated Georgia's ratification of key international treaties related to the management of oil spills.[29] In recent years, the Group's International Development Association has also partnered with the US Millennium Challenge Corporation (MCC) to fund and oversee the rehabilitation of Georgia's Soviet-era gas pipelines and to help structure a United Nations Development Programme–sponsored Carbon Facility so that Georgia can benefit from carbon financing.

Above and beyond specific pipeline projects, the World Bank has been a central player in Georgia's larger energy infrastructure and transport sectors. Through the Transport Rehabilitation Project, the Group has aided the government in reforming Georgia's railways, a key mode of energy transport from Baku to Georgia's Black Sea ports. The World Bank is one of the only development organizations apart from the MCC, with whom Georgia signed a Compact Agreement in 2005, involved in helping the country rebuild its dilapidated highway network. The Roads Project is aimed at upgrading and improving major sections of the East-West Highway, augmenting road-maintenance programs, and exploring promising public-private partnerships to attract other sources of financing through the Bank's Infrastructure Pre-Investment Facility. In 2005, only an estimated 5 percent of Georgia's road network supported the transit economy and foreign trade. According to the Bank, "the long transit times, despite the relatively short distance and poor road conditions, add to transport costs and deter usage of the Georgian transit route." Highway rehabilitation remains one of the key modes of exploiting ever-higher transit volumes arriving from Central Asia and the Caucasus.[30] In parallel with these efforts, the IFC helped finance construction of new terminals at the Tbilisi and Batumi airports.[31]

Georgia's energy infrastructure is among the worst in the former Soviet Union and the unreliability of the country's power supply and inadequate gas transmission systems continue to inhibit private-sector growth. To address these challenges, the World Bank has begun in recent years to work with the governments of Japan and Germany, the EBRD, and the United States Agency for International Development (USAID) to upgrade the electricity sector and gas networks in the country.[32]

Crisis Response

Since the Rose Revolution, the government of Georgia has made remarkable progress in reforming outmoded and inefficient Soviet-style governance practices. The World Bank ranked Georgia fifteenth out of 181 economies in its annual *Doing Business* report in 2009; the Bank's annual reports compile a set of globally recognized benchmarks that in combination measure the relative ease of doing business in a country.[33] As a testament to the magnitude of recent progress, when Georgia embarked on a major set of reforms in 2004–5 it was ranked 112th. Despite such progress, the global economic crisis and the August 2008 war with Russia acted as a set of twin shocks to the Georgian economy; their combined effects led to a 3.2 percent contraction of GDP in the second half of 2008 and prompted the need for a $750 million Standby Arrangement with the IMF to avert a potential balance-of-payments crisis.

The World Bank played a key coordination role alongside other Western partners in managing the crisis response to the August 2008 conflict. The Bank and the United Nations Development Group collaborated on a Joint Needs Assessment (JNA) methodology to determine the extent of the damage and the amount of financial assistance necessary for rehabilitation.[34] Building upon these efforts, the World Bank cohosted a donors conference with the European Union in October 2008, wherein thirty-eight countries and fifteen organizations pledged an estimated $4.5 billion in international assistance over the period of 2008–10 to help Georgia rebuild and meet urgent postconflict humanitarian needs. To help channel and manage a portion of these new funds, the government of Georgia asked the World Bank to assist with the establishment and administration of a multidonor trust fund called the Phoenix Fund. The Phoenix Fund adheres to World Bank standards for financial management and procurement and is intended to attract debt and equity financing for large infrastructure projects.[35] The crisis illustrated not only the breadth of the World Bank's convening power, but its larger integrative effect with respect to other Western organizations.[36]

Among the more serious consequences of the 2008 conflict has been the impact on long-term investor confidence in Georgia's potential as a transit route for transcontinental energy flows and trade. Although direct damage to energy infrastructure as a result of the conflict was relatively modest, the temporary suspension of oil shipments and gas supply interruptions that resulted underscored both Georgia's strategic geographic location and its attendant vulnerabilities. To prevent further such obstacles to the realization of thriving energy and transport links among Azerbaijan, Georgia, and Turkey, the Bank has begun to address some of these new urgencies that have surfaced as a result of the Russian incursion—the need for greater security of energy and infrastructure via improvements in the reliability of power transmission and gas storage facilities to cushion potential interruptions of imports in the future.[37] Above and beyond dealing with short-term supply disruptions, the Group concluded in its JNA, "Georgia needs to enhance its energy security by continuing to develop domestic energy resources."[38] Under the International Bank for Reconstruction and Development–enclave Energy Project, the Group has funded a feasibility study for the development of a hydropower resource at Khudoni that could significantly improve the security of Georgia's energy supply. The Group estimates such a project upon completion could generate an estimated 20 percent of current hydropower production and would be structured "as an export-oriented sale of power to neighboring countries" and thereby "enhance Georgia's role in the regional energy trade."[39]

Building Institutional Capacity

Unlike Azerbaijan, corruption is much less of a hindrance to growth and development in Georgia. However, significant lack of capacity means that in terms of supporting energy- and transport-sector integration with its neighbors, there is ample opportunity for achieving even greater impact. As of January 2008, the IFC's portfolio in Georgia stood at an estimated $93 million.[40] The World Bank Group has supported national reform of major institutions such as the judiciary and facilitated reductions in barriers to commercial activity and the privatization of key enterprises. The Judicial Reform Project provides for the "attestation of judges; publication of laws, decrees, and court decisions; and establishment of a training center for judges." The Enterprise Rehabilitation Project is "helping strengthen management of privatized enterprises as well as pre-privatization restructuring of state-owned enterprises."[41] One of the IFC's main advisory projects in Georgia is the Business Enabling Environment Project, which involves conducting surveys of small and medium-sized enterprises and providing guidance on licenses, permits, and inspections.[42]

The Group also works closely with other institutions that complement the World Bank's efforts to expand institutional capacity and secure adequate capital financing, such as USAID, the EU, and the EBRD. To help mitigate the effects of the global credit crunch, the IFC recently announced it would partner with the EBRD and the US Overseas Private Investment Corporation to provide the Bank of Georgia, the largest and most systemically important bank in the country, with enough capital to continue lending to Georgian companies and small enterprises. The IFC also helped establish the ProCredit Bank of Georgia, the country's first bank to specialize in the provision of capital to micro and small enterprises. Lastly, the Group facilitated the post–Rose Revolution government's implementation of a decentralization project to increase local government capacity throughout the country in accordance with the European Charter on Local Self-Governance.[43]

TURKEY

Energy Sector and Risk Mitigation

The World Bank Group's assistance to Turkey has been broad-based and diversified across many sectors, aimed at supporting Turkey's development as a subregional anchor in the Caucasus, and as a major middle-income partner of the Bank and a globally significant emerging market.[44] In past years, the country has been among the largest borrowers in the Bank's Europe and Central Asia Division and Turkey presently represents the IFC's fifth-largest exposure after the BRIC countries (Brazil, Russia, India, and China).[45] As of September 2007, the IFC's committed portfolio in Turkey totaled $1.6 billion and the IFC is committing a similar amount during the 2008–2011 budgeting period.[46] In terms of promoting Turkey's crucial strategic and geographic role in the formation of transit routes through the Azerbaijan-Georgia-Turkey subregion to the West, the World Bank envisages substantial future Bank Group financing in the energy sector, "which reflects the importance of energy security and energy efficiency." The World Bank Group's Multilateral Investment Guarantee Agency, founded to promote foreign direct investment and provide political risk guarantees to potential investors, is very active in Turkey and helps support infrastructure projects.[47] Overall, the Bank's Country Partnership Strategy for Turkey projected up to $6.2 billion in new commitments from the Bank for the years 2008–2011.[48]

The Group financed the initial feasibility study for the Turkish portion of the BTC route terminating at Ceyhan, and its early support for the transnational pipeline project played an important role in securing the requisite diplomatic heft and financial backing from other extra-regional actors and private-sector companies for its completion. The IFC's investment in the BTC project consisted of a loan of up to $125 million for its own account ($76 million for the portion of the pipeline built in Turkey) and another $125 million loan in commercial bank syndication.[49] According to a senior World Bank official, the Group also held discussions with the Turkish government "about complying with good international environmental standards, safeguard issues, resettlement of peoples, [and] dealing with the Kurds" during BTC and South Caucasus pipeline construction. Other examples of Bank-sponsored projects that promote and emphasize common environmental standards throughout the region include the Caspian Oil and Gas Dialogue and the Strategic Partnership on the Black Sea and Danube Basin, a joint effort between the World Bank and relevant UN agencies to address environmental degradation of these crucial water systems.

The World Bank is likewise committed to gas sector development in Turkey over the next few years, to increase the "reliability and stability of gas supply in Turkey by implementing critically needed gas storage and network infrastructure, and strengthen the Turkish Pipeline Cooperation in its operations as a financially stable and commercially managed corporation."[50] In light of the August 2008 explosion along the BTC pipeline, for which the breakaway Kurdistan People's Party took responsibility, and the similar route traversed by the South Caucasus gas pipeline, such World Bank–administered projects could help mitigate future gas supply disruptions by providing underground storage facilities should political and ethnic tensions result in sabotage along the energy and transport linkages among Azerbaijan, Georgia, and Turkey. Lastly, Turkey's unique vulnerability to natural disasters—particularly earthquakes—is of related concern to the safety and security of its transit pipelines. The BTC pipeline alone travels through seven active earthquake faults in Turkey, and between 1992 and 2008 Turkey experienced 130 earthquakes of magnitude 5.0 or higher on the Richter scale.[51] The Bank supports disaster preparedness through a variety of different avenues, including the World Bank Institute's Disaster Risk Management Program and the Bank's Istanbul Seismic Risk Mitigation and Emergency Preparedness Project.[52]

COORDINATION WITH OTHER EXTRA-REGIONAL ACTORS

The World Bank Group partners with a number of international financial institutions, bilateral donor governments, foundations, and private-sector corporations to promote regional integration, development, and good governance on multiple levels. As a result of its considerable resources, near-universal membership, and global reputation, the World Bank is in a singular position to bring other organizational actors with specialized technical expertise and funding together to address the full range of development needs for Azerbaijan, Georgia, and Turkey and thereby support the Western integrative process. The Group has worked alongside efforts of the EBRD, the IMF, DFID, USAID, the US Trade and Development Agency (USTDA), and the US Department of the Treasury, as well as the European Union via its European Neighborhood Policy (ENP). As in other member countries, the World Bank and the IMF country teams regularly coordinate their policy advice, particularly in such overlapping areas as tax policy and financial and banking sector reform. More specifically, the two organizations collaborate on the Financial Sector Assessment Program (FSAP) reports of member countries, a joint World Bank and IMF effort launched in 1999 to promote the soundness of financial systems worldwide.[53]

Azerbaijan

In Azerbaijan, the World Bank has taken the lead on donor coordination in the energy and transport sectors. Beginning with the Azerbaijan Highway Project, the EBRD and Asian Development Bank have come to an agreement with the Group to supplement its technical assistance program in large-scale infrastructure and transport. The Group works with the Islamic Development Bank to strengthen capacity building in various areas, and with the EU's TACIS program on taxation modernization. The USTDA and the EBRD were critical partners working with the Group on SOCAR restructuring, and the UNDP worked with the Bank on integrating the Millennium Development Goals into the government's official poverty reduction program. The US Treasury and the World Bank have jointly endeavored to modernize the budget preparation process, and the Group has worked alongside USAID on utility regulatory reform. The Bank also relied on Canadian- and Swiss-sponsored Consultant Trust Funds to enhance programs that support improved financial reporting across government ministries, regulatory agencies, and private-sector corporations in Azerbaijan. Regarding judicial reform, the Bank is one among many organizations working with the Ministry of Justice and the courts system on rule-of-law reform, training of judges, and the improvement of legal education. In this arena, the Bank's IDA collaborates with the American Bar Association, the International Committee of the Red Cross, and the Soros and Eurasia foundations.[54]

Georgia

In Georgia, the Group has played a key role coordinating numerous extra-regional efforts and donors. In the energy sector, the World Bank took the lead in bringing together a number of key partners including USAID, the EBRD, and KfW, a German government–owned development bank.[55] MCC and the World Bank have worked together in Georgia to improve municipal services and infrastructure support programs. The Georgia Donor Framework Group, which aims to ensure coordinated development assistance, is facilitated by the World Bank and the UNDP and includes the IMF, the European Union, the United States, the United

Kingdom, France, Germany, Sweden, and the Netherlands as members. The World Bank Group works closely alongside a number of US-sponsored projects in Georgia, as the country ranks among the top world states for American aid per capita (in the fiscal years 1992–2006 Congress approved an estimated $1.76 billion in total funds) and the United States is Georgia's largest bilateral donor.[56] The Group has also collaborated on several EU-sponsored projects under the ENP. Through its IFC development arm, the World Bank is uniquely positioned to seamlessly integrate the efforts of the public and private sectors. The IFC has worked closely with BP and its oil and gas partners on the Regional Development Initiative (RDI), discussed in chapter 5, which is designed to bring socioeconomic benefits to Azerbaijan, Georgia, and Turkey. The RDI constitutes a number of discrete projects in all three countries and includes the Georgia Corporate Governance Project, which brings together the IFC, the Canadian International Development Agency, and BP and partners to improve corporate-governance practices of Georgian companies and banks.[57] The Public Sector Management Reform Support Project represents a pooled funding arrangement between the Bank, DFID, the Netherlands, and the Swedish International Development Cooperation Agency to improve public sector management in Georgia.[58]

Turkey

In addition to its extensive collaboration on EU-related initiatives in Turkey, the Group has partnered with the IMF and Germany's GTZ international development organization in implementing municipal and public sector reform. As it does in Azerbaijan and Georgia, the Bank works with the IMF in Turkey to facilitate financial sector reform, collaborate on regular FSAP reports, and contribute to macroeconomic monitoring. The IMF and the World Bank also work together in the arena of social security reform, particularly as social inclusion and adequate safety nets for the poor are key aspects of the Copenhagen Criteria for EU accession.[59] To further improve and monitor the business climate in Turkey, the Bank has been a driving force behind the creation of the International Advisory Council, a group comprised of key members of the Turkish cabinet, CEOs of major companies, Turkish business association heads, and senior representatives of international financial institutions that convene regularly to advise the government on future priorities for reform.[60] The Council of Europe Development Bank and the European Investment Bank work together with the Bank on earthquake risk mitigation, and ecological and natural disaster preparedness, as well as environmental and infrastructure projects.[61] The UN and the World Bank have also identified areas where stronger collaboration in Turkey is a priority, particularly in the arenas of poverty alleviation and local development.[62]

WORLD BANK GROUP RANKINGS

Senior World Bank officials have highlighted the effect of World Bank Group rating and ranking reports on integrative reforms in Azerbaijan, Georgia, and Turkey. Highly publicized and universally recognized rankings such as the World Bank–IFC annual *Doing Business* report (DB) can motivate reform in various member countries, particularly as these sets of benchmarks are closely monitored by investors and donor governments. Although the World Bank administers multiple ranking reports across various categories, *Doing Business* is among the best known globally and ranks 178 countries on ten core dimensions of business regulation. The Azerbaijani government has in recent years begun to make a concerted effort to improve its traditionally low scores, and in 2009 the country was named *Doing Business*'s top regulatory reformer worldwide. The country's overall rank for the ease of doing business jumped from ninety-seventh place to thirty-third between 2007 and 2008, the biggest one-year jump in *Doing Business* history.[63] Georgia was named the top reformer among the Commonwealth of Independent States in 2006 and listed among the top ten reformers globally that same year and in 2008. During his time in office as Georgia's prime minister, Zurab Noghaideli publicly cited the desire to improve the country's World Bank rankings as an impetus for implementing further economic reform. A 2008 Independent Evaluation Group (IEG) report affirmed that policy makers across many different countries tend to use *Doing Business* rankings as a "tool" in developing policy reforms.[64] Furthermore, the DB rankings can have a multiplier effect when implemented and promoted by other development aid agencies. The MCC uses DB rankings as a basis for resource allocation, which continues to add to the prestige and press attention associated with this particular set of indicators. Ultimately, the IEG analysis concluded "the DB exercise has demonstrated that a cross-country ranking exercise can be effective in spurring dialogue and motivating interest and action."[65]

The World Bank Institute also publishes its annual *Governance Matters* index, the most comprehensive set of indicators measuring various components of governance available,[66] and the Group collaborates with the EBRD on a set of Business Environment and Enterprise Performance Surveys (BEEPS) that includes all the countries of Central and Eastern Europe, the former Soviet Union, and Turkey. According to the BEEPS sponsors, "the objective of the survey is to provide feedback from enterprises on the state of the private sector in countries of operation by assessing the constraints to private-sector growth and creating statistically significant business environment indicators that are comparable across countries through interviews with firms in the manufacturing and services sectors."[67] Each of these respective sets of rankings and surveys includes Azerbaijan, Georgia, and Turkey and provides common incentives for their harmonization of standards and integration in the energy and transport sectors.

CONCLUSION

The World Bank Group would most likely be a critical development player in Azerbaijan, Georgia, and Turkey even if their energy and transport sectors were not integrating. But, the role of a TERA in this case has provided focus and a coordinating mechanism around those sectors which has provided impetus for their integration. Because the World Bank's commitment to an infrastructure project is a signal to the private sector about the legitimacy of a proposed initiative (the so-called crowding-in effect), and because of the lucrative nature of many of these energy and transport infrastructure initiatives, the governments of Azerbaijan, Georgia, and Turkey have been accepting of the same World Bank Group–mandated governance and rule-of-law reforms that are necessarily tied to its financial support. This harmonization of standards allows for integration. Had international financial institutions not been key facilitators of such major transnational projects, it is not only unlikely they would have been completed in a timely manner—if at all—it is highly unlikely they would have been accompanied by a much broader and more consequential set of integrative measures. These achievements are compounded by the coordinating role that the World Bank Group continues to play with other extra-regional actors in the region, a role that has made the work of other international institutions all the more effective in driving integration among Azerbaijan, Georgia, and Turkey's energy and transport sectors.

NOTES

1. Alexandros Petersen, "Consolidating EU Energy Security: The East-West Transport Corridor and Western Integration," *Romanian Journal of International and Regional Studies* 4, no. 2–3 (2008): 20–23. See also Petersen, "Integrating Azerbaijan, Georgia, and Turkey with the West: The Case of the East-West Transport Corridor," CSIS Commentary, September 10, 2007.

2. International Finance Corporation, *Sustainability Report 2004* (Washington, DC: World Bank Group, 2004). See part 5, "Challenges of Development: The BTC Pipeline and Extractive industries Review," 32–35.

3. IFC Extractive Industries Review Reports, available at http://go.worldbank.org/T1VB5JCV61.

4. Anthony Cholst (lead country officer, South Caucasus Country Management Unit, Europe and Central Asia Region, World Bank), interview with author, August, 10, 2007.

5. World Bank, *Azerbaijan: Country Partnership Strategy FY07–10* (Washington, DC: World Bank, 2006), 27.

6. International Finance Corporation, *Sustainability Report 2004*, 32–35.

7. International Finance Corporation, "New IFC and Microfinance Bank of Azerbaijan Initiative to Provide Local Suppliers with Greater Access to Finance," news release, May 16, 2007.

8. International Monetary Fund, *Republic of Azerbaijan: 2008 Article IV Consultation; Staff Report* (Washington, DC: International Monetary Fund, 2008); see "Statement by Thomas Moser, Executive Director for Republic of Azerbaijan, and Martin Lanz, Advisor to Executive Director."

9. Christian E. Petersen and Nina Budina, "Governance Framework of Oil Funds: The Case of Azerbaijan and Kazakhstan" (Washington, DC: World Bank, 2003), 4–5.

10. World Bank, *Azerbaijan: Issues and Options Associated with Energy Sector Reform* (Washington, DC: World Bank, 2005), 16.

11. Ibid., 4.

12. Ibid., 34.

13. Coffey International Development, *Validation of the Extractive Industries Transparency Initiative (EITI) in the Republic of Azerbaijan* (Berkshire, UK: Coffey International Development, 2009).

14. World Bank, *Azerbaijan: Issues and Options Associated with Energy Sector Reform,* 11.

15. Mahmoud Ghafouri, "The Caspian Sea: Rivalry and Cooperation," *Middle East Policy* 15, no. 2 (2008): 82.

16. World Bank, *Azerbaijan: Country Partnership Strategy FY07–10,* 12.

17. World Bank, *Azerbaijan: Building Competitiveness; An Integrated Non-Oil Trade and Investment Strategy (INOTIS),* vol. 1, *Summary Report* (Washington, DC: World Bank, 2003), 2.

18. International Monetary Fund, *Republic of Azerbaijan: Staff Report for the 2007 Article IV Consultation* (Washington, DC: International Monetary Fund, 2007), see annex 2, pp. 5–6.

19. World Bank, *Azerbaijan: Country Partnership Strategy FY07–10,* 5–8.

20. Ibid., 25.

21. Transparency International, "Corruption Perceptions Index 2008," www.transparency.org.

22. International Monetary Fund, *Republic of Azerbaijan: 2008 Article IV Consultation; Staff Report,* 5.

23. World Bank, *Azerbaijan: Country Partnership Strategy FY07–10,* 22.

24. Ibid., 30.

25. World Bank, *Georgia: Country Partnership Strategy* (Washington, DC: World Bank, 2005); see executive summary 5.

26. International Monetary Fund, *Georgia: 2009 Article IV Consultation and Second Review Under the Stand-By Arrangement; Staff Report* (Washington, DC: International Monetary Fund, 2009), 7.

27. International Finance Corporation, "Azerbaijan and Georgia: Early Oil Project."

28. World Bank, *Georgia: Country Partnership Strategy,* 60.

29. International Finance Corporation, "Case Studies: Georgia Oil Institution Building Project"

30. World Bank, *Georgia: Country Partnership Strategy,* 23.

31. World Bank, *Georgia: Country Partnership Strategy Progress Report for the Period FY06–FY09* (Washington, DC: World Bank, 2008), 5.

32. World Bank, *Georgia: Country Partnership Strategy,* 22–23.

33. World Bank, *Doing Business 2009 Europe and Central Asia: Comparing Regulation in 181 Economies* (Washington, DC: World Bank, 2008).

34. United Nations and World Bank, *Georgia: Summary of Joint Needs Assessment Findings Prepared for the Donors' Conference of October 22, 2008 in Brussels* (Washington, DC: United Nations and World Bank, 2008).

35. Ibid., 29.

36. World Bank, *Georgia—World trade indicators 2009: Trade brief* (Washington DC : World Bank, 2009) http://documents.worldbank.org/curated/en/2009/12/17013370/georgia-world-trade-indicators-2009-vol-1-2-trade-brief.
37. United Nations and World Bank, *Georgia: Summary of Joint Needs Assessment Findings*, 11.
38. Ibid., 26.
39. World Bank, *Georgia: Country Partnership Strategy*, 23.
40. World Bank, *Country Partnership Strategy Progress Report for Georgia for the Period FY06–FY09*, 8.
41. World Bank, *Georgia: Country Partnership Strategy*, 22.
42. World Bank, *Country Partnership Strategy Progress Report for Georgia for the Period FY06–FY09*, 9.
43. International Monetary Fund, *Georgia: 2009 Article IV Consultation and Second Review under the Stand-By Arrangement; Staff Report*, 9.
44. World Bank, *Turkey: Country Partnership Strategy for the Period FY 2008–2011* (Washington, DC: World Bank 2008), 1.
45. Ibid., 20.
46. Ibid., 25.
47. Ibid., 21.
48. Ibid., 38.
49. Ibid., see annex 2.
50. World Bank, *Status of Projects in Execution Report for FY2008* (Washington, DC: World Bank Group, 2008), 1014.
51. World Bank, *Turkey: Country Partnership Strategy for the Period FY 2008–2011*, 44.
52. Ibid., 34.
53. International Monetary Fund, *Republic of Azerbaijan: 2008 Article IV Consultation; Staff Report*, 5–6. See also World Bank, *Georgia: Country Partnership Strategy Progress Report for the Period FY06–FY09*, 9.
54. World Bank, *Azerbaijan: Country Partnership Strategy FY07–10*; see annex 6, pp. 100–103.
55. World Bank, *Georgia: Country Partnership Strategy*, 80.
56. Jim Nichol, "Georgia: Recent Developments and U.S. Interests," CRS Report for Congress (Washington, DC: Congressional Research Service, 2008).
57. World Bank, *Georgia: Country Partnership Strategy Progress Report for the Period FY06–FY09*, 9.
58. Ibid.
59. World Bank, *Turkey: Country Partnership Strategy for the Period FY 2008–2011*, 37.
60. Ibid., p. 20.
61. Ibid., 38.
62. Ibid., 38.
63. World Bank, *Doing Business 2009 Europe and Central Asia.*
64. World Bank Independent Evaluation Group, *Doing Business: An Independent Evaluation; Taking the Measure of the World Bank—IFC Doing Business Indicators* (Washington, DC: World Bank 2008), see executive summary xxv. The Independent Evaluation Group is an independent unit within the World Bank that monitors its activities, identifies lessons learned from Bank projects, and is meant to serve as an objective assessor of the Bank's efforts and outcomes. The IEG reports directly to the Bank's Board of Executive Directors.
65. Ibid., 53.
66. Robert I. Rotberg, "Strengthening Governance: Ranking Countries Would Help," *Washington Quarterly* 28, no. 1 (2004): 72–73.
67. See World Bank, "Business Environment and Enterprise Performance Surveys," http://data.worldbank.org/data-catalog/BEEPS.

FOUR

Transnational Extra-Regional Actors: TRACECA and INOGATE

Two main EU-pioneered programs, the Transport Corridor Europe-Caucasus-Asia (TRACECA) and Interstate Oil and Gas Transport to Europe (INOGATE), emerged in the 1990s to integrate the EU, Eastern Europe, the Caucasus, Turkey, and Central Asia. In so doing, they served as tools with which the EU, the transnational extra-regional actor in this case, was able to contribute to the integration of Azerbaijan, Georgia, and Turkey in the sectors of transport and energy. This chapter will examine the activities of TRACECA and INOGATE, in their full scope, as well as in Azerbaijan, Georgia, and Turkey. A clear example of how integration is achieved in the transport and energy sectors through the involvement of TERAs should emerge. It should also become apparent that integration in the transport sector is due to a variety of processes, taking from security communities theory, neofunctionalism, and neoliberal institutionalism, whereas integration in the energy sector is a neoliberal institutionalist dynamic.

TRACECA

As part of the European Union's Technical Aid to the Commonwealth of Independent States (TACIS) program, TRACECA is rooted in a May 3, 1993, conference in Brussels, where representatives of the EU and ministers of transport and commerce from Kazakhstan, Kyrgyzstan, Tajikistan, Uzbekistan, Armenia, Azerbaijan, and Georgia (Turkey joined the project later) signed an agreement to develop an EU-funded "transport corridor on the West-East direction from Europe."[1] The corridor sought to develop "an East-West axis from Central Asia, across the Caspian Sea, through the Caucasus, across the Black Sea to Europe."[2] A sharp economic decline and serious backlogs in out-of-date transport systems after the fall of the Soviet Union brought the governments of the TRACECA-participating countries and the European Commission together as they sought more active incorporation of post-Soviet markets into "the international economy by improving and enhancing trade and transport."[3] INOGATE's integrative objectives were similar: facilitating secure energy supplies to the EU through the updating of post-Soviet energy infrastructure and dissemination of best practices for its operation. The TRACECA and INOGATE programs seized on the opportunity for participant countries to expand their economic ties with Europe and integrate more closely among themselves in the transport and energy spheres.

TRACECA's transport routes pass through territories rich with raw materials and energy resources in need of modern functional outlets for delivery to Western consumers. Not surprisingly, a significant portion of the transported goods are "oil and oil products, minerals, metals, construction materials, chemical products, and consumer goods."[4] Accelerated development of oil and gas fields in Azerbaijan, Kazakhstan, and Turkmenistan since the end of the Cold War is the key contributing factor to the constant increase in the volume of hydrocarbon cargo transits via TRACECA routes. According to a TRACECA representative, between 2005 and 2007, sea transport through the Caspian alone accounted for more than 13.5 billion tons of oil and oil products, with transit volumes increasing tenfold since 1996.[5] In 2006, the volume of trade among TRACECA member countries rose 22 percent and reached close to €300 billion, while trade with the EU accounted for €200 billion.[6] Furthermore, the proximity of Central Asian countries to China's booming economy makes them an attractive region to serve as a transit conduit to the West. This is a particularly important point in view of the increasing trade turnover between Asia and Europe, which exceeds $2 billion, while the transport expenses make up approximately $200 billion.[7] One of the main visions behind TRACECA was that the project would serve as a stabilizing factor in Eurasia by stimulating the economies of its participants and integrating them into the global economy and among themselves.

Indeed, given newly opened borders in the wake of the Soviet Union's collapse, the geopolitical location of the TRACECA participants, and enhanced trade between Europe and Asia, "the rise of freight flows on the one hand and logical reverse freight flows on the other hand," the development of transport and communications links between the two regions made simple economic sense.[8] Thus far, TRACECA has focused on "trade facilitation and harmonization of the legal basis, with inter-modal transport [as well as] the transport modes such as road, railway, civil aviation, and maritime transport."[9] After its central multilateral agreement was signed in September 1998 in Baku, TRACECA has progressed from one project to another: sixty-one technical assistance and fifteen investment projects between 1996 and 2006, developed within its Action Program framework, in agreement with TACIS rules and regulations.[10] As Demchenko and Kaminsky put it:

> Considering the TRACECA route in the context of developing trade relations between Europe and Asia and location of main commodity producers in Asia and consumers in Europe on the one hand, as well as the flow of goods of potential major cargo owners on the other hand, it can be concluded that the delivery of goods to Europe through TRACECA looks much more attractive than other alternative routes.[11]

With thirteen members and three observer countries potentially interested in joining the program, TRACECA's expansion may eventually include Afghanistan. TRACECA's secretariat also plans for links with the EU's Trans-European Networks (TEN) program, a similar integrative mechanism that emerged in the late 1980s to link "various regions and national networks [in Eurasia] by modern and efficient infrastructure."[12]

The EU has financed TRACECA with an annual allocation of €9 to 11 million. However, since 2004, the participating countries have been encouraged to contribute 15 percent to the general budget, with the EU continuing to provide financial assistance on a smaller scale.[13] EU institutions have now concentrated their efforts on "consulting and investment projects for the TRACECA program."[14] Brussels has laid out the main directions of TRACECA's strategic development until 2015, including the formation of modern infrastructure, cooperation with Trans-European Networks, guarantee of naval and aviation security, as well as energy transit routes. As TRACECA's mission has expanded over the years, investment in its projects from international financial institutions and European private investors has increased.[15]

The Nuts and Bolts of TRACECA

TRACECA is at the core of the main trade route linking Europe and Asia. The idea to restore the ancient Silk Road, linking East-West trade and transport, can be traced to the September 1990 Vladivostok International Conference when then–Soviet Minister of Foreign Affairs Eduard Shevardnadze (later president of independent Georgia) voiced his support for more open collaboration "between the countries of Eurasia."[16] The idea was favorably received by the countries of the Caucasus and the EU, resulting in the formation of TRACECA in May 1993 in Brussels. Its initial members included Armenia, Azerbaijan, Georgia, Kazakhstan, Kyrgyzstan, Tajikistan, Turkmenistan, and Uzbekistan. In 1996, Ukraine became a full member of TRACECA, while Moldova was granted membership in 1997 and Bulgaria, Romania, and Turkey in 2000. In 2002, Afghanistan, China, Greece, and Iran expressed interest in joining TRACECA, while several other countries participated in the project as observers.[17] With Bulgaria and Romania's accession to the EU in 2007, TRACECA became inextricably tied to EU institutions.[18] Given its mandate, TRACECA's goals are perhaps unsurprisingly ambitious: strengthening the political and economic sovereignty of CIS countries in the TRACECA region to enable their effective participation in the world economy; supporting regional cooperation between TRACECA countries; promoting regional investment by international and private financial resources; and linking the Eurasian transport corridor with European and world transport systems.[19]

I. TRACECA's ultimate objective is to extend a railway from the Chinese port of Lianyungang along the Yellow Sea to the Georgian seaports of Poti and Batumi along the Black Sea, as well as to Turkey's railroad network and the Marmara tunnel under the Bosporus, which will then access Western Europe. Designed to promote closer collaboration and dialogue among participant countries, TRACECA has aimed to sustain transit fees at a competitive level and to professionalize procedures at transborder crossings along internationally standardized lines.

The program's participating countries have concluded agreements on the transport of large cargoes via the TRACECA route, demonstrating that the "route is the shortest and potentially the fastest route from Central Asia to deep-water ports linked to world markets."[20] Realization of this potential has been a key driving force for TRACECA partner states to attract investors from the public and private sectors. That said, transnational extra-regional actors such as the EU and international financial institutions, as well as the integrative effects their involvement brings, have been central to most of TRACECA's projects. As Abid Sharifov, vice-premier of Azerbaijan, noted at an anniversary conference of TRACECA in Baku on December 4, 2008, up to that date, the European Commission had provided funding of $157 million for TRACECA's sixty-one technical and fifteen investment projects.[21] In fact, the EU's central support has provided the reputational backing needed for other international organizations to participate in eliminating non-geographic barriers to transport goods: TRACECA has helped to attract large investments from development partners including the European Bank for Reconstruction and Development, which has committed funds for capital projects on ports, railways, and roads along the TRACECA route; the World Bank, which has financed new capital projects on roads in Armenia and Georgia; the Asian Development Bank which has allocated substantial funds for road and railway improvement: and the Islamic Development Bank, which has invested in the development of the transport sector in the TRACECA countries. The EU is supporting the program with other European Commission projects to further enhance regional cooperation and economic sustainability in the region, such as the Southern Ring Air Routes project and the Oil and Gas Pipeline project (INOGATE).[22] According to the program's estimates, international financial institutions had invested over €1.3 billion into various TRACECA projects through 2008.[23]

TRACECA's program portfolio has grown significantly since its inception. Its ongoing projects include: development of coordinated national transport prices; improvement of maritime links between TRACECA corridors and Trans-European Network corridors; creation of perishable goods licensing and certification centers; training of operational air transport staff in South Ring states; analysis and forecasting of traffic flows for TRACECA countries and interregional transport integration, including organization of the transport dialogue between the EU and NIS in the framework of the European Neighborhood Policy; strengthening of transport training capacity in post-Soviet countries; and development of the port at Aktau, on Kazakhstan's Caspian coast, including a master plan and feasibility study.[24]

II. Although TRACECA projects have faced various challenges in their implementation stages as well as inherent issues with the program's weak overall strategy, they have been relatively successful. New roads and railways are under construction along the entire span of the TRACECA corridor.[25] According to Svante Cornell, research director of the Central Asia–Caucasus Institute at Johns Hopkins University, the EU's new Eastern European members are interested in injecting renewed momentum into the program.[26]

III. Construction of ports, bridges, and other transport infrastructure is underway in conjunction with the renewal of old roads, railways, bridges, and ports.[27] Investment projects to build terminals in the ports of Ilyichevsk, Poti, Batumi, Turkmenbashi, Baku, and Aktau have already been implemented, opening regular railway and ferry communication lines along Varna-Ilyichevsk-Poti-Batumi and Baku-Aktau, tying Azerbaijan and Georgia together.[28] The construction of a railway tunnel under the Bosporus, which would connect the continents of Europe and Asia across the straits, was to be finalized in 2009.[29] According to a TRACECA representative, the average annual turnover of goods for the past five years along the route increased by 10 to 15 percent.[30] In 2005, about fifty million tons of goods were transported from Asia through the Caucasus to Europe and vice versa, while thirty million tons of goods transited through the same route within the seven months of 2008.[31]

IV. The TRACECA secretariat plans to develop the Samarkand-Termez highway as an integral part of the trans-Afghan corridor, which was solidified by an agreement between Uzbekistan, Iran, and Afghanistan. The modernization of the Kungrad-Beineu-Aktau railway as well as the construction of the Andijan-Kashgar railway via Kyrgyzstan has received support from TRACECA.[32] Even Tajikistan, Central Asia's most remote country, is bound to benefit from the EU-backed TRACECA. The country's "transcontinental highway Tashkent-Dushanbe or Termez-Dushanbe passing through the territory of Gorny Badakhshan via the Kulma passing with access to the Karakum highway allows it to reach China and then Japan and the Asian-Pacific region, as well as Pakistan and India."[33] While isolated Turkmenistan has until recently ignored TRACECA, burgeoning ties with the EU and Western countries at the diplomatic level may lead to eventual participation in the program. All of these projects in Central Asia would have to connect to Europe through the Caucasus and Asia Minor, which in turn further integrates the transport infrastructure of Azerbaijan, Georgia, and Turkey.

V. Azerbaijan Railways upgraded its "control system in line with the TRACECA project [in 1998–99 and] an oil depot and new station were constructed in 1998 in Diubendi to handle transport of domestic and Kazakhstan petroleum."[34] TRACECA seeks to improve 60 percent of Kazakhstan's railway tracks. Modernization of the country's Caspian Aktau port has received $54 million from the European Bank for Reconstruction and Development with the capacity "to handle 2 million tons of dry cargoes and 6 million tons of oil."[35] The same year, Ukraine opened the Poti-Ilyichevsk transport center and "the completed rehabilitation of 1870 kilometers of track infrastructure."[36] Railway modernization in other member countries also received significant funding from the EU and international financial institutions.

VI. Since 2000, the volume of cargo transit along the TRACECA route has increased between six to ten times, depending on the area. According to the general secretary of TRACECA's Permanent Secretariat of the Inter-Governmental Commission, Rustan Jenalinov, turnover of goods over the past five years has reached on average 10 to 15 percent, with hydrocarbons making up 70 percent of that figure.[37] Jenalinov is convinced that the volumes of cargoes transiting through the TRACECA corridor are bound to increase with increased transport integration between China and Western Europe, especially given that the annual growth of turnover of goods among TRACECA member states exceeds 20 percent. Moreover, China's booming economy elevates the role of the region as a key transit hub. China accounts for 50 percent of trade turnover between Asia and Europe and by 2013 occupied the first place in trade turnover in the world, making the Caucasus and Turkey a vital global transit route.[38] Ever since the trans-Caucasian trunk railway took effect on May 13, 1996, cargo transport across the TRACECA region has grown steadily.[39]

VII. At a May 2007 meeting of TRACECA's Inter-Governmental Commission, which seeks to unify member country efforts, "centered on the formation of sustainable, effective and integrated multimodal transport systems," participants adopted a strategy to run through 2015.[40] According to Jenalinov, the strategy's main priorities until 2015 are "provision of security and safety of freights, [which requires] strengthening of institutional and legislative bases as well as increasing the potential of the transport personnel, and improvement of transport infrastructure via defining viable projects, optimal use of investments and raising them from various sources."[41] It is worth pointing out that the strategy gives a special role to forwarding agents and operators of cargoes as they optimize a complex process.[42] Such an endeavor will incorporate "harmonization of laws, usage of modern logistical technologies, strengthening of nonstate regulations, and increasing professional standards [including] training of cargo forwarding agents in ten countries, [which] is currently in the final stages."[43]

VIII. The key component of TRACECA's efforts so far has been simplification and unification of laws to streamline cross-border trade. An accord to regulate "transit traffic signed by Georgia, Azerbaijan, Turkmenistan and Uzbekistan in 1996 in Sarakhs [Turkmenistan] and joined later by Turkey goes a long way towards coordinating railway business activities. It was the first example of legislative changes towards promoting the project."[44] Integration and harmonization of transport legislation in the region is a continuous process. The program's presence in the region is evident as there have been "frequent regional conferences and seminars, close partnerships with international financial institutions, TRACECA's technical assistance and investment projects, and an increasing number of those using the corridor, particularly in the Caucasus."[45]

IX. A document entitled "Visa TRACECA" has been in the works for several years. Its adoption will allow for the lowering of various transport-related duties and tariffs to increase the competitiveness of goods transported via TRACECA. Currently, various duties levied on a car going from across the Caspian to Turkey via Azerbaijan and Georgia reach up to $400 just to cross the borders. As a result, the price of the goods increases significantly. The incentive for TRACECA member countries to adopt integrative measures such as Visa TRACECA is clear.[46]

TRACECA's Evolution

X. Since its inception in 1993, the TRACECA program has grown significantly with EU support. That said, because TRACECA was established in an atmosphere of haste, a common strategy among the participating countries to identify priorities has not been developed in detail. As admitted by a TRACECA representative, the program's potential has yet to be fully met.[47] Coordination of actions to develop and implement transport, tariff, customs, and tax policies among TRACECA member countries as well as procedures to cross borders and development of transport infrastructure remain open for discussion.[48] According to the EU's 1998 evaluation of TRACECA projects, "any notion of strategy appears to have been superimposed after the decisions have been taken. There is no evidence that there has been a focused strategy that has guided decisions. This is particularly evident at the present time when most of the technical assistance projects have come to an end, with varying degrees of success."[49]

XI. Some of the criticisms lodged at the EU regarding TRACECA centered on its steadily fading interest in the program and its lack of a guiding vision. According to some reports, EU officials appeared increasingly unenthusiastic about these programs and tried to subsume regional transport issues to the EU-Russia dialogue.[50] According to Bakyt Kakchekeev, Kyrgyzstan country director for the consulting firm Task Force Innovations International, TRACECA has lost its identity among other organizations that also seek to improve transport routes, such as the Eurasian Economic Community, the Asian Development Bank transport initiative, the Eurasian Continental Bridge, and others.[51] In Kakchekeev's view, TRACECA's progress was not particularly distinguishable from that of other transport-related organizations.[52] Similarly, TRACECA participant countries have also been criticized for absence of interest in advancing the program after the EU passed on the mantle, particularly as some countries raised transport duties, while other key countries in the region, such as Turkmenistan, have still not joined.[53]

XII. That said, despite the challenging political and economic environment in many of TRACECA's member countries, the program has been moderately successful in Azerbaijan, Georgia, and Turkey and has contributed to the integration of their transport sectors. The difficulty in measuring the success of the projects has been mostly due to the lack of comprehensible norms and standards, as TRACECA's initial goals have changed and "are not sufficiently definitive to make measuring the effectiveness of the program easy."[54]

XIII. As the initial goal of the TRACECA program was to boost the efficacy of the transport sector, it involved fact finding about conditions in the transport sphere, provision of practical solutions, and professional training. These efforts produced mixed results, as some projects were more successful (for example, the Policy and Regulatory Framework projects) than others (such as the Traffic Forecasting and Dolphin projects).[55] Technical assistance from the EU was key to sustaining many projects, which raised a question about the continuing viability of TRACECA once the EU reduced its funding.

XIV. The EU continues to grant financial assistance to TRACECA through the TACIS program. Up to 2006, TACIS funded "62 technical assistance projects worth €105 million and 14 investment projects worth €52 million."[56] However, since early 2000, the EU began reducing its funding to the program as TRACECA's member states assumed a greater financial role to support the activities of the Permanent Secretariat, its working groups, and relevant operations. The gradual transfer of responsibilities has allowed TRACECA member countries to become more involved in advancing the projects. This has partially been a product of greater integration of their transport sectors.[57] According to EU evaluations, TRACECA has succeeded from its early stages in "demonstrating the potential for regional cooperation in transport and has achieved a significant level of awareness at the political level."[58] With EU support, TRACECA was able to improve the quality of transport services and created a foundation for future actions as well as "visibility for TACIS and for the EU assistance to the region generally."[59]

XV. The gradual handover of TRACECA from the EU to the program's member states has continued its development. Realizing the program's promising future, the TRACECA participants continue to hold regular discussions on the improvement of the transport corridor and its integration with other corridors. They have also pursued concrete measures to implement investment projects and raise money to fund them. When Kazakhstan took over TRACECA's chairmanship in 2007, the country's minister of transportation, Serik Akhmetov, stressed Astana's commitment to "integration initiatives in the transportation sphere."[60]

XVI. Member countries have also made significant individual investments in some projects. For example, Azerbaijan "invested $200 million in development of sea transport and seven new tankers have been bought through these means. Some $500 million has been invested in development of air transport."[61] Further, the member states have also exerted effort to increase training and professional development of staff in their transport sectors. One such project, launched in 2008, seeks to improve the preparation and training of students in the transport field in higher education institutions within the framework of TRACECA.[62] The program's participating countries are reportedly determined to keep the program running and they are exploring ways to find more funding.

INOGATE

The Interstate Oil and Gas Transport to Europe program, or INOGATE, was established in 1995 as TRACECA's counterpart, with a more narrow focus on the development of oil and gas transport and transit systems. With its Technical Secretariat based in Kyiv, which is the official representation of the program, and a regional office in Tbilisi, INOGATE's members comprise twenty-one countries, including countries of the former Soviet Union and Turkey. The EU and several international financial institutions and international organizations are the program's observers.[63]

As noted earlier, similar to TRACECA, INOGATE has sought to promote regional integration in the Caucasus and Asia Minor via its technical assistance by connecting the transport infrastructure in those regions with that of EU member states. The ultimate goal of the program is to facilitate "secure energy supply for the EU . . . while acting as a catalyst in attracting private investors and international financial organizations to finance such projects."[64] INOGATE hopes to improve pipeline systems and transport of oil and gas, aiming at restoring and modernizing the energy infrastructure in the program's participating states, providing technical training to staff, and evaluating the feasibility of different transport routes for energy from the Caucasus and Central Asia to Europe. The program is considered an important tool to improve "the safety and security of energy supply in both INOGATE participating countries and the EU" as well as a platform for collaboration and discourse.[65] It is well recognized in Europe and was well received by its twenty-one member states. From its inception, the program has reached out to garner investment in order to enhance energy supply networks and has served as a conduit for bringing together private investors and international financial institutions for energy infrastructure projects.

Formed as INOGATE's institutional pillar, the Umbrella Agreement on the Institutional Framework for the Establishment of Interstate Oil and Gas Transportation Systems, which took effect in February 2001, is termed

> an interstate agreement that sets out an institutional system designed to rationalise and facilitate the development of interstate oil and gas transportation systems and to attract the investments necessary for their construction and operation. INOGATE intervened at a moment when all the interstate agreements reached under Soviet times were cancelled and no legal agreement or rules for co-operation existed between the newly-independent states.[66]

The Umbrella Agreement, which seeks to facilitate the development of energy transport systems and investments needed for their construction, calls for more efficient construction and use of transport systems tying the Caucasus and Asia Minor to Central Asia and the EU on either side and seeks to help reduce international risks in cross-border oil and gas pipeline projects. All twenty-one countries are signatories to this agreement. As stated in a TACIS report evaluating the INOGATE program,

> there is evidence that INOGATE has contributed to regional cooperation, especially through the Umbrella Agreement. Through this Agreement, INOGATE has helped to replace a command and control model of development within the region based in central planning by a market model based in business law, contract negotiation and common technical standards.[67]

INOGATE has also coordinated its actions with other "initiatives such as the EU-Russia Energy Dialogue, the Euro-Mediterranean Energy Forum (the 'Barcelona Process', which includes the southern Mediterranean countries), and the EU Energy Community."[68] It was predicted in 2007 that INOGATE's technical assistance program would be incorporated into the European Neighborhood and Partnership Policy, although such a change has not been officially promulgated yet.

Similar to TRACECA, INOGATE's track record has been mixed. The program has gained support from high-profile political circles in its participating countries and has served as a platform for dialogue and collaboration. Further, INOGATE has played a role in prefeasibility and feasibility studies of new energy transport routes, particularly those passing through the Caspian Sea to Central and Eastern Europe. It also seeks to give "advice on technical, financial and legal issues related to project preparation."[69] One of the immediate ways that the INOGATE program has assisted its participating states is via funding of important small investment projects in "hazardous interstate infrastructures, based on the need to improve the safety, security and transparency of the gas networks," under the Umbrella Agreement.[70]

Serving as a platform for greater collaboration, the program's progress in gaining recognition and attracting investor attention has met some success. According to INOGATE's website,

> over the period of 2003–2004, the INOGATE Technical Secretariat carried out the first coordinated investment attraction activities of the INOGATE Programme. The Investment Attraction Campaign culminated in an Investment Round-table on March 31, 2004, where 12 projects were presented to international financial institutions and potential private investors. Interest was expressed in several projects by financial representatives and currently follow-up actions are in progress.[71]

The EU has played a key role in INOGATE's projects by way of financing studies and "preparing applications for project finance from international financial institutions, training of staff, designing the regulatory framework for creating a sustainable regional and export market, and implementation of small-scale investments in hazardous cross-border infrastructure (such as pipelines) as demonstration projects for interstate transport."[72] However, as has been noted above, the EU has been steadily reducing its funding to both TRACECA and INOGATE and encouraging greater financial contribution from participating countries. It is expected that the member states of both programs will take greater responsibility for them. In the case of INOGATE, they will increase their ownership of it by providing more contributions and "together with other regional energy co-operation programmes, INOGATE . . . will form an important pillar."[73]

Although the ideas behind the INOGATE and TRACECA programs have been sound and served as grounds for mutually beneficial economic integration between the member states via streamlining key transport sectors, the implementation phase has at times been difficult, and the outcomes of many projects have not been as effective as expected. While TRACECA appears to have made noticeable progress in improving some transport routes and increasing cross-border cargo transit, coupled with greater recognition of its goals and projects in the region, INOGATE is at least a few steps behind its transport counterpart. Most of INOGATE's efforts have concentrated on feasibility studies and evaluations of projects, while the actual implementation of most of its projects has received less-than-favorable assessments, particularly from one of its main funding sources, TACIS. INOGATE has had little effect on some of its members, such as Kyrgyzstan, for example, which is tied by bilateral deals on energy products with Kazakhstan and Russia, but in Azerbaijan, Georgia, and Turkey, a number of energy projects have been successfully completed under the aegis of INOGATE.[74] According to Svante Cornell, the main issue with INOGATE is that it was never built to be an actor, that is to say, it had no real structure from the beginning and its purpose was to provide technical assistance, while having no mandate to design and implement a comprehensive set of projects, which would require significant political will.[75] Frederick Starr shares these views, stating that INOGATE has been quite passive, especially in Central Asia, but that it has contributed to regional integration in the energy sector among Azerbaijan, Georgia, and Turkey due to their proximity to the EU and importance in the so-called East-West or Southern Energy Corridor.[76]

INOGATE's technical assistance focused on four areas—"market-related studies, resource evaluation studies, project feasibility studies and institutional studies"—all aimed at standardization across all member states.[77] As the TACIS assessment of INOGATE's activities in 1997 indicates, the program began supporting regional integration via technical assistance of institutional development, which has helped "to turn potential projects into realistic investments, identifying relevant finance for these projects, carrying out financial analysis assessing business risk, and raising the required finance."[78] Importantly, the parties involved in INOGATE supported the implementation of small-size projects, with the potential to add to the more efficient use of pipelines and other energy infrastructure. The 1998 and 1999 programs covered issue areas such as

> preliminary engineering studies for oil and gas transport across the Caspian Sea; diagnostic and preliminary engineering studies for the renovation of critical oil pipelines identified by the 1996 audit; preliminary engineering studies for the renovation of critical gas pipelines identified by the 1996 audit; further priority investments in gas and oil infrastructure.[79]

INOGATE has executed its projects in two phases: the feasibility studies phase from 1996 to 2000 and the implementation phase since 2004. Some projects, assessed in the TACIS report issued in the year 2000, were not effective, nor had they met initial goals as planned. It is worth examining in greater detail which projects proved ineffective. The Regional Oil and Gas Management Project, launched in 1994 and designed to help the countries to handle their energy transmission systems, has garnered attention of senior officials but proved ineffective in accomplishing many of its initial goals.[80] The Feasibility Study for Oil and Gas Pipelines through the Caspian Sea, aimed at evaluating the feasibility of building additional energy transit routes via the Caspian, met a similar fate as it was evaluated as inefficient.[81] The Rehabilitation of Crude Oil and Oil Products Transportation Networks project, seeking to restore, "modernize and rationalize oil and oil pipeline projects in the countries participating in the INOGATE program" and to implement an audit of major crude oil routes to Europe, was assessed by TACIS as "being poorly performed."[82]

As specified in the TACIS evaluation, other projects have received low scores, including the Rehabilitation, Modernization, and Rationalization of Existing Gas Transmission Networks—Audit of Existing Infrastructure Central Asian Gas Transmission System (Kazakhstan, Kyrgyzstan, Tajikistan, Turkmenistan, Uzbekistan); Rehabilitation, Modernization, and Rationalization of Existing Gas Transmission Networks—Audit of Existing Infrastructure in Caucasus; Rehabilitation, Modernization, and Rationalization of Existing Gas Transmission Networks—Audit of Existing Infrastructure in Belarus, Moldova, and Ukraine; and Rehabilitation, Modernization, and Rationalization of Existing Gas Transmission Networks—Audit of Existing Infrastructure from Central Asia to the Caucasus, Ukraine, Moldova, and Belarus. Although the overall efficiency of these projects was satisfactory, they were mostly ineffective, except for their training elements.[83] The Institutional Issues and Strengthening of Regional Cooperation project was ranked as more successful in terms of performance, cost-effectiveness and management, and having an impact on Baku-Tbilisi-Ceyhan negotiations.[84] Criticisms of INOGATE have further focused on the necessity to develop a more focused institutional structure for regional collaboration instead of ad hoc meetings.

INOGATE's track record is, however, not altogether negative, and there has been some significant progress in its projects completed between 2004 and 2007. These include

> development of maintenance excellence in NIS gas companies; emergency priority actions for security and improvement of exploitation conditions for gas transport systems in Central Asia; feasibility study for a satellite monitoring and accident prevention system for the security of gas transport and storage infrastructures; financing small-scale investments for regional oil and gas infrastructure in three Caucasus countries, Turkey and Central Asia; gas pipelines diagnostic mobile laboratory; IFI Technical Assistance Support Fund for facilitating investments in oil and gas projects; gas security of supply to Armenia in the framework of the closure of the Mezamor Nuclear Power Plant; pre-feasibility studies and facilitation for developing the North-South; reform of Ukraine Gas Transit Network operations Phase III; supply of IT equipment and training for development of maintenance practices in NIS gas companies; supply of a pipeline network information systems to the Central Asia gas companies.[85]

In its most recent efforts to engage the INOGATE participants and improve the environmental conditions surrounding the exploitation of oil and gas resources in the region, the EU announced in December 2008 the launching of a "two-year project to deliver improvements to the environmental impact of oil and gas activities in Central Asia."[86] As part of INOGATE, the $3.63 million project would be a key player in "developing a sustainable and secure energy supply between Europe and the Central Asian region," between which Azerbaijan, Georgia, and Turkey serve as a bridge.[87] The EU will provide technical assistance to monitor and control the environmental influence of oil and gas transport operations. As a transnational extra-regional actor, the EU's push to link its energy networks with those of Central Asia has led to regional integration among the countries geographically in between in the Caucasus and Asia Minor.

Despite the fact that INOGATE's projects by and large have proven less effective than hoped, the overall approach of INOGATE to serve as a catalyst of integration of energy transit routes has assisted in building functional links between the former Soviet states and Europe, with the Caucasus and Turkey as a major conduit in that process. Looking back at the problems that INOGATE has faced, such as the lack of leadership to steer its goals and mission, lack of proper coordination of actions and duplication of efforts, poor training, and an attendant lack of investment, assessment of its role as an integrative mechanism should center on how it could improve to further its original goals of integrating the energy transport of the post-Soviet states, with an eye on the real impact of the overall program.

In a clear reference to regional integration, the EU's assessment of INOGATE pointed out that the program "has an important political dimension" and is recognized by high-level government officials in the region.[88] Further, the program has encouraged an active political dialogue among the participating countries and the EU. International energy companies are also familiar with this dialogue. But one of the main accomplishments of INOGATE is that it has successfully endorsed "the ideas of international business law and project assessment among senior officials and politicians, creating a common language of debate and negotiation; the small investment projects have been generally welcomed by beneficiaries and often cited as concrete achievements on environmental issues."[89] Its future success largely hinges on better leadership, clearer vision and achievability of goals, and better program management.

CONCLUSION

Both the TRACECA and INOGATE programs have been necessary components of the larger EU strategy to engage Azerbaijan, Georgia, and

Turkey. Although they have had a mixed track record of success, their presence has been an additional impetus to regional efforts on integration in the transport and energy spheres. The efforts to connect the transport routes of Eurasia with Europe have borne some fruit, as evidenced by increased willingness to cooperate on mutually beneficial economic matters. Both the TRACECA and INOGATE programs have provided first-hand experience to the countries of the Caucasus and Asia Minor on how to operate more closely, by lowering economic barriers such as tariffs, opening borders, pursuing mutually beneficial goals, and capitalizing on trade opportunities.

TRACECA in particular has proven successful compared to INOGATE as the cargo transit and trade turnover between TRACECA-participating countries has steadily increased over the years. TRACECA is much better recognized in the region than INOGATE, largely due to its track record of tangible impact on the development of transport routes. TRACECA's executive secretary of the Permanent Secretariat in 2003, Ambassador Zviad Kvatchantiradze, emphasized at the time that the growing volume of cargo transit via TRACECA is clear evidence of the program's overall success.[90] According to the general secretary of TRACECA's Permanent Secretariat of the Inter-Governmental Commission, Rustan Jenalinov, in 2005 about fifty million tons of cargo were transited through the countries of the South Caucasus from Asia to Europe and vice versa, while it was close to fifty-four million tons in 2007.[91] As a result of these activities, new projects and plans to attract more investment have accentuated TRACECA's need in the region. Svante Cornell and Hikmet Hadjy-zadeh consider both TRACECA and INOGATE important and promising programs.[92] According to Cornell, as the need for these programs is still there and the member states have put forth a vision and financial support for various projects, the missing link is political will and better coordination of efforts.[93] However, Bakyt Kakchekeev holds that the future of both programs is vague and that it is likely they may not accomplish much in their current state, while continuing to maintain their formal existence.[94]

While its accomplishments are evident and the participating countries realize the potential for further collaboration, TRACECA has yet to develop a focused strategy, boost its funding capability, and improve coordination of joint actions. As Jenalinov points out, "differences in economic development and in technological advances of the member states' infrastructures, as well as an interest of other countries in the corridor, require attendant institutional changes."[95] These efforts will also depend on further collaboration with international organizations and international financial institutions. In addition, a regional information exchange system among the TRACECA as well as INOGATE member states will be essential to increased harmonization of efforts.[96] The transport development strategy in effect through 2015, which envisions the boosting of transit capabilities of TRACECA members, will be a further litmus test of the program's success as an integrative mechanism. While INOGATE has clearly facilitated integration among Azerbaijan, Georgia, and Turkey, the program's viability is less measurable compared to TRACECA, due to unclear implementation of its goals, as well as because of comparatively decreased recognition among business and government circles in the participating countries. Both programs, however, serve as illustrative example of how a TERA, the EU, operates on the ground to foster integration.

NOTES

1. TRACECA website, http://www.traceca-org.org.
2. "EU/NIS: Commission Approves TACIS Assistance Projects for Former USSR," *European Report* (Brussels), September 27, 1997.
3. TRACECA website, http://www.traceca-org.org.
4. Rustan Jenalinov (general secretary of TRACECA's Permanent Secretariat of the Inter-Governmental Commission), "Koridor, rozhdenyi dlia konkurentsii" [Corridor, born to be competitive], interview with Olga Skripkina, RZD Partner, November 7, 2008, http://www.rzd-partner.ru/interview/2008/11/07/333180.html.
5. Anar Ismail (TRACECA representative), interview with the author, February 5, 2009.
6. Ibid.
7. Victoria Panfilova, "'Viza TRASEKA' pozvolit snizit' transportnye sbory i poshliny" ["Visa TRACECA" will facilitate the reduction of transportation duties and tariffs], *Ekho* (Azerbaijan), March 17, 2008. http://www.ng.ru/courier/2008-03-17/17_silkway.html
8. TRACECA website, http://www.traceca-org.org.
9. Ibid.
10. Ibid.
11. I. O. Demchenko and P. D. Kaminsky, "TRASEKA kak tsel' economicheskogo sotrudnichestva" [TRACECA as a goal of economic collaboration], Donetsk National University of Economics and Trade, 2008. http://www.rusnauka.com/28_NIOXXI_2008/Economics/35420.doc.htm.
12. Under the terms of chapter XV of the treaty (articles 154, 155, and 156), the European Union must aim to promote the development of Trans-European Networks as a key element for the creation of the internal market and the reinforcement of economic and social cohesion. This development includes the interconnection and interoperability of national networks as well as access to such networks. Trans-European Networks website, http://ec.europa.eu/ten/index_en.html.
13. Panfilova, "Viza TRASEKA" [Visa TRACECA].
14. TRACECA website, http://www.traceca-org.org.
15. Ibid.
16. Teimuraz Gorshkov and George Bagaturia, "TRACECA: Restoration of Silk Route," *Japan Railway and Transport Review* 28 (2001): 52.
17. Asian Development Bank, "Obzor TRASEKA i dostizheniya proekta garmonizatsii protsedur peresecheniia granits" [Review of TRACECA and the achievements of the project's harmonization of border-crossing procedures], report for the Regional Seminar on Trade Facilitation and Customs Modernization, Issyk-Kul, Kyrgyzstan, August 4–8, 2003.
18. Ismail, interview with the author.
19. Gorshkov and Bagaturia, "TRACECA: Restoration of Silk Route," 53.
20. Asian Development Bank, "Obzor TRASEKA" [Review of TRACECA].

21. "Azerbaijan Develops Transport Infrastructure under TRACECA," *Trend*, December 4, 2008.
22. TRACECA website, http://www.traceca-org.org.
23. Rustan Jenalinov, "TRASEKA otkryt dlia vsekh?" [TRACECA is open to everyone?], interview with Anna Nezhinskaya, RZD Partner, November 9, 2007, http://www.rzd-partner.ru/interview/2007/09/11/310945.html.
24. TRACECA website, http://www.traceca-org.org.
25. Ibid.
26. Svante Cornell, interview with the author, February 4, 2009.
27. TRACECA website, http://www.traceca-org.org.
28. Panfilova, "Viza TRASEKA" [Visa TRACECA].
29. Ibid. The tunnel finally opened in 2013 after a delay of four years and is partially operational.
30. Ismail, interview with the author.
31. Ibid.
32. Panfilova, "Viza TRASEKA" [Visa TRACECA].
33. Ibid.
34. Gorshkov and Bagaturia, "TRACECA: Restoration of Silk Route," 54.
35. Ibid.
36. Ibid.
37. Jenalinov, "Koridor, rozhdenyi dlia konkurentsii" [Corridor, born to be competitive], interview with Olga Skripkina.
38. Jenalinov, "TRASEKA otkriyt dlya vsyekh?" [TRACECA is open to everyone?], interview with Anna Nezhinskaya.
39. Viktor Abaturov, "Nazad—v budushchee, ili v poiskakh evraziiskogo transportnogo konsensusa" [Back—to the future, or in search of the Eurasian transport consensus]. *Economicheskoe Obozrenie* no. 7 (2003) http://www.centrasia.ru/newsA.php?st=1068242160.
40. A. A. Denisenko (deputy general secretary of KazATO), "Problemy ispol'zovaniia mezhdunarodnogo koridora TRASEKA," presentation at the "State and Private Sector Partnership to Solving Transportation Problem of Countries with No Access to Sea" roundtable, Batumi, Georgia, May 19, 2007.
41. Jenalinov, "TRASEKA otkryt dlia vsekh?" [TRACECA is open to everyone?], interview with Anna Nezhinskaya.
42. Ibid.
43. Ibid.
44. Gorshkov and Bagaturia, "TRACECA: Restoration of Silk Route," 53.
45. Asian Development Bank, "Obzor TRASEKA" [Review of TRACECA].
46. Panfilova, "Viza TRASEKA" [Visa TRACECA].
47. Ismail, interview with the author.
48. Ibid.
49. TACIS Evaluation Unit, "Evaluation of the Tacis Interstate TRACECA Programme" (Brussels, 1998), 41, available at http://ec.europa.eu/europeaid/how/evaluation/evaluation_reports/reports/tacis/951421_en.pdf.
50. Vladimir Socor, "Too Little, But Not Yet Too Late," *Wall Street Journal*, October 6, 2003.
51. Bakyt Kakchekeev, interview with the author, February 2, 2009.
52. Ibid.
53. "Azeri Official Believes Europe-Caucasus-Asia Corridor is Victim of Indifference," *BBC Monitoring*, September 22, 2000.
54. TACIS Evaluation Unit, "Evaluation of the Tacis Interstate TRACECA Programme," 41.
55. Ibid., 44.
56. European Union-TACIS Program, "Caucasus/Central Asia," document available at: http://www.kugm.gov.tr/BLSM_WIYS/TRACECA/tr/projeler/devam_eden/Strengthening_of_Transport/20100506_150909_10426_1_64.doc.

57. Ibid.
58. "Evaluation of the TRACECA Interstate Program – ref. 951421," http://ec.europa.eu/europeaid/how/evaluation/evaluation_reports/evinfo/1998/951421_ev_en.pdf.
59. Ibid.
60. "V ramkakh proekta TRASEKA Kazakhstan planiruyet realizovat' okolo 80 proektov na $30 milliardov" [Kazakhstan plans to implement about 80 projects for $30 billion within the TRACECA project], *Panorama* (Kazakhstan), December 14, 2007.
61. "Azerbaijan Develops Transport Infrastructure under TRACECA."
62. "TRASEKA nachala realizatsiiu novogo proekta v stranakh SNG" [TRACECA began implementing a new project in NIS], Kazinform, October 20, 2008.
63. Alexandros Petersen, "Integrating Azerbaijan, Georgia, and Turkey with the West: The Case of the East-West Transport Corridor," *CSIS Commentary*, September 10, 2007, 14.
64. INOGATE, "In brief," no date, http://www.inogate.org/pages/1?lang=en.
65. INOGATE website, http://www.inogate.org.
66. Ibid.
67. Environmental Resource Management, *Evaluation of Tacis Inter-State Energy and INOGATE Programmes and Related Actions Implemented in the Framework of National Programmes* (Brussels: European Commission, 2000), 1:57.
68. INOGATE website, http://www.inogate.org.
69. Ibid.
70. Ibid.
71. Ibid.
72. "EU/NIS: INOGATE Summit Opens Energy Umbrella Agreement over NIS," *European Report* (Brussels), July 24, 1999.
73. INOGATE website, http://www.inogate.org.
74. Kakchekeev, interview with the author.
75. Cornell, interview with the author.
76. Frederick Starr (chairman, Central Asia–Caucasus Institute, Johns Hopkins University), interview with the author, February 10, 2009.
77. Environmental Resource Management, *Evaluation of TACIS Inter-State Energy and INOGATE Programmes*, 1:F3.
78. Ibid., 61.
79. Environmental Resource Management, *Evaluation of TACIS Inter-State Energy and INOGATE Programmes*, 1:63.
80. Ibid., 1:45.
81. Ibid.
82. Ibid., 1:47.
83. Ibid., 1:48–51.
84. Ibid., 1:51.
85. INOGATE, "Overview of INOGATE Projects (1996–2008)," 2008, http://www.inogate.org/projects?collection=1996_2008&lang=en.
86. "Caspian Basin: EU Funds Environmental Protection Program for Caspian Basin," *EurasiaNet*, December 19, 2008.
87. Ibid.
88. Environmental Resource Management, *Evaluation of TACIS Inter-State Energy and INOGATE Programmes*, 1:69.
89. "Evaluation of Tacis Inter-State Energy and INOGATE Programmes and Related Actions implemented in the Framework of National Programmes – ref. 951599," http://ec.europa.eu/europeaid/how/evaluation/evaluation_reports/evinfo/2000/951599_ev_en.pdf.
90. G. Kulieva, Yubiley TRACECA: Pervye itogi novogo 'Shelkogo puti'" [TRACECA's jubilee: First results of the new "silk road"], *Ekho* (Azerbaijan), June 29, 2003.
91. Jenalinov, "Koridor, rozhdenyi dlia konkurentsii" [Corridor, born to be competitive], interview with Olga Skripkina.

92. Cornell, interview with the author, and Hikmet Hadjy-zadeh (vice president, FAR Center [Azerbaijan]), interview with the author, February 3, 2009.
93. Cornell, interview with the author.
94. Kakchekeev, interview with the author.
95. Jenalinov, "Koridor, rozhdenyi dlia konkurentsii" [Corridor, born to be competitive], interview with Olga Skripkina.
96. Ibid.

FIVE

The China–Central Asia Pipeline: A Counterfactual on the Role of TERAs

Azerbaijan, Georgia, and Turkey are becoming increasingly intertwined with one another to the exclusion of their neighbors. This pattern of integration has two overriding characteristics. First, the integration that has occurred so far seems to be limited to the most highly regulated, discrete, and mobile sectors, such as energy and transport, without spilling over into far more fundamental areas such as the political and security spheres. Second, the integration that has occurred has been, in some sense, a top-down process, or even cultivated by supranational bodies—transnational extra-regional actors—such as the European Union and international financial institutions. Emblematic of that integration are the major highway projects and oil pipelines that run through all three countries. The process is symbiotic, and each of the three countries benefits more from cooperation than they would otherwise. Each of the three countries had something to contribute to the Baku-Tbilisi-Ceyhan oil pipeline, for instance: Azerbaijan possessed a high-value natural resource; Georgia possessed the necessary transit corridor through a relatively stable and democratizing ally; and Turkey possessed access to world markets, especially the European Union.

Azerbaijan, Georgia, and Turkey each at least formally commit to the European integration project and benefit from the European Union's Neighbourhood Policy, an agenda of policy reforms directed at the countries on the EU periphery. In this package of policy reforms is significant technical assistance in the energy and transport sectors through programs such as INOGATE (oil and gas) and TRACECA (transport), which aimed at rehabilitating existing infrastructure and harmonizing regulatory and administrative frameworks among the member countries. Standardization would improve predictability, which itself would improve the attractiveness of the region to foreign investment. The combination of the European Union's Neighbourhood Policy toward Georgia and membership accession negotiations toward Turkey have helped improve cooperation between the two countries on taxation, trade, customs, rail and sea transport, and other issues. The European Union is by no means the only TERA at work in Azerbaijan, Georgia, and Turkey: the World Bank, the International Finance Corporation, the Asian Development Bank, the Organization for Security and Cooperation in Europe, and a host of others have been actively providing financing and technical assistance to projects in Azerbaijan, Georgia, and Turkey with broader, overarching agendas in mind.

While other external actors such as Russia have attempted to widen a sphere of influence in Azerbaijan, Georgia, and Turkey, not all external actors operate as TERAs. This point deserves some emphasis. Russia could in theory engineer an integrative process in the region, attempting to harmonize the regulatory and legal frameworks of Azerbaijan, Georgia, and Turkey with its own, and as a consequence exert more influence or even control over the energy and transport sectors of the three countries. What is happening with regard to the European Union is qualitatively different. The EU is only partially engaged in the integrative process in order to reap material rewards such as a stable natural gas supply. The EU agenda is much broader than this, and seeks to export norms of democratization, human rights, economic development, and the rule of law to countries on the periphery. TERAs, unlike other external actors such as Russia, are spurring a process of integration in Azerbaijan, Georgia, and Turkey that is bringing the three countries closer together *among themselves* to the exclusion of their neighbors. In other words, while Russian-led integration, for instance, could technically lead to harmonization of administrative and regulatory frameworks in Azerbaijan, Georgia, and Turkey (hegemonic integration, as Keohane recognized), integration would occur only secondarily and in one direction. Azerbaijan, Georgia, and Turkey would each be integrating with Russia and only indirectly with each other even though a region-wide harmonization process would be taking place.

This point is subtle, but it has important implications for a counterfactual case study, which can help in more specifically isolating TERA involvement in infrastructure projects as an explanatory variable for why integration has occurred—at least in the energy and transport sectors—around these projects in Azerbaijan, Georgia, and Turkey. The creation of a pipeline from Turkmenistan through Uzbekistan and Kazakhstan and on to the Xinjiang Province in the People's Republic of China bears certain similarities to the pipelines through Azerbaijan, Georgia, and Turkey running from Baku through Tbilisi to Ceyhan (BTC) and Erzurum (BTE). First, the geographic motivations behind the pipelines are similar: a bypass of existing Russian avenues for landlocked resources to reach world markets. Second, the pipeline traverses several international borders and requires sophisticated contractual, regulatory, and taxation agreements among those countries, as well as a mechanism for dispute resolution. Finally, and not insignificantly, like the oil and gas pipelines in Azerbaijan, Georgia, and Turkey, the so-called China–Central Asia gas pipeline is extremely large, traversing nearly 1,200 miles at a cost of over $7 billion.[1] Nonetheless, the natural gas pipeline through Central Asia lacks a central feature of the BTC and BTE pipelines: the participation of TERAs. While the natural gas pipeline in Central Asia is driven by China and may place some pressure on Turkmenistan, Uzbekistan, and Kazakhstan to bargain in good faith to reach agreement on tariffs and fees, customs and pipeline security, China does not possess the broader agenda of TERAs active in Azerbaijan, Georgia, and Turkey. In addition, while Turkmenistan, Uzbekistan, and Kazakhstan all reached agreement on contract terms with China, they are not integrating *among one another* to the exclusion of their neighbors. The example of the China–Central Asia pipeline allows for an answer to the question: would a major pipeline project through three countries like Azerbaijan, Georgia, and Turkey contribute to the integration of their energy sectors were it not for the involvement of TERAs in its development?

THE CHINA–CENTRAL ASIA PIPELINE AND ENERGY IN CENTRAL ASIA

China, a country with a rapidly diversifying energy portfolio, has naturally seen the producer states of Central Asia as a partial solution to its energy needs. Since 2006, Xinjiang, an autonomous region of China bordering Kazakhstan and Kyrgyzstan, has become a leading producer of natural and oil-associated gas. In 2007, Xinjiang produced 21.2 billion cubic meters (bcm) of gas, up from less than 5 bcm in 2002, and is on target to produce 30 bcm in 2010; likewise, the region's share in China's total domestic gas production rose from less than 3 percent in 1990 to 31 percent in 2007. With proven reserves at 1.3 trillion cubic meters (tcm), it holds a quarter of China's total gas reserves. China planned to have increased the share of gas in its energy balance to 5.3 percent by 2010, with gas consumption expected to reach 140 bcm per year with at least 90 bcm produced domestically, 20 bcm imported by pipeline, and 30 bcm as liquefied natural gas. Domestically, China is building a series of west-to-east pipelines from producing areas to the south and east of the country. In 2004, a four-thousand-kilometer-long pipeline started delivering gas from Xinjiang to ten provinces, and in 2007, construction was approved on a second pipeline to stretch five thousand kilometers.[2] The Chinese government anticipates a 7 percent annual growth rate in natural gas use between 2000 and 2020. Total energy use will double in that period, with natural gas fueling much of the growth. China has also built an extensive network of internal pipelines, stretching twenty-two thousand miles by 2010.[3]

In line with China's rapid economic growth, the country's energy imports have doubled in the past five years. As Central Asia is comparatively close to rapidly developing Xinjiang, it is logical that China would seek to participate in the sharp global competition over Central Asian hydrocarbon resources. China imports about ten to twenty million tons of oil from Kazakhstan every year, which is more than 10 percent of China's total oil imports, and operates a pipeline running from Atyrau in western Kazakhstan to Alashankou in Xinjiang region, which connects to the Kenjiyak-Atyrau pipeline. Even though the volume of energy imports from Central Asia to China may not be strategically significant yet, it is growing at a rapid rate and projections are more than optimistic. Besides being the largest economy of Central Asia, Kazakhstan has proven oil reserves of 2.7 billion barrels. In 1997, China pledged $10 billion in investment for oil exploration and construction of infrastructure. In December 2005, China bought Kazakhstan's flagship oil company PetroKazakhstan for $4.8 billion, and also invested $700 million in building a pipeline from Kazakhstan to China via Kyrgyzstan.[4]

Observers have dramatically referred to China's growing interest in Central Asian energy and the international competition that has followed as a new "Great Game" and a new "Silk Road."[5] By 2007, intense competition in Central Asia existed between three gas pipeline projects: (1) expansion of the pipelines to Russia; (2) the trans-Asia gas pipeline to China; and (3) the United States– and European Union–sponsored trans-Caspian pipeline. It is unclear whether there is enough Central Asian gas for all three proposed routes. Russia aims to modernize the Central Asia–Center pipeline with assistance from both Kaztransgaz and Gazprom, and assist with construction of a Caspian littoral pipeline, which could cost up to $1 billion. Provisional agreements between Kazakhstan, the Chinese government, and the state-owned Chinese National Petroleum Company (CNPC) set out a provisional agreement that would consist of two trunk pipelines: first, the Kazakh portion of the Turkmenistan-China gas pipeline, running through southern Kazakhstan and primarily transporting Turkmen gas, and second, a pipeline from resource-rich western Kazakhstan to western China. The latter proposed pipeline could be an important interconnector by linking Kazakhstan's main gas routes into a unified system and provide flexibility in exports by linking the Central Asia Center and the Bukhara-Ural gas trunk pipelines.[6] The growing rivalry among Russia, China, and the European Union via a trans-Caspian pipeline has boosted the bargaining position of Turkmenistan, Kazakhstan, and Uzbekistan.

As with Kazakhstan, China is also providing energy assistance to Uzbekistan, and Uzbek President Islam Karimov visited China in May 2006 and yielded Beijing's first serious pledge of investment to the Uzbek energy sector. China is helping develop the modest oil fields in the Fergana valley and both countries signed a framework agreement worth $1.5 billion in July 2005. The two countries have since signed an agreement for a $950 million long-term loan and an additional $350 million soft loan. Although China's oil needs are expected to grow, from 6.3 billion barrels per day to 10 billion in the next twenty years, oil's share of total energy consumption will decline as natural gas increases, at a rate of 7.8 percent per year.[7] In general, however, Chinese energy companies have expressed less interest in Uzbek natural gas because the country's oil potential is dwarfed by Kazakhstan's and its gas potential by Turkmenistan's.

Chinese outreach to Central Asia in fields other than oil and gas is almost as significant. China is responsible for building a vast network of rail and road links through Kyrgyzstan, a transit state, in order to connect China and Uzbekistan. In 2005, China pledged an investment of $900 million in Kyrgyzstan for telecommunications infrastructure, which is almost half of the country's GDP. While Uzbekistan is not a major exporter of oil, the country sits on vast reserves of uranium, which has drawn attention not only from China but also from India and even Russia, whose reserves of uranium have declined.[8] As a secondary player in the gas game, Uzbekistan has sought to maximize returns from its strategic geographical location for Turkmen gas to be delivered to China. Moreover, if new gas deposits are found in Uzbekistan, they could be linked to a Turkmenistan-China pipeline.[9]

The ultimate center for natural gas exploration and production, however, and the laser-like focus of China's Central Asian outreach, is Turkmenistan. In a search for export routes around Russia, and with a path through Iran blocked by Western donors, Turkmenistan's President Niyazov visited China in April 2006 to negotiate a contract for a pipeline stretching deep into the Chinese interior under which Turkmenistan would deliver to China 30 bcm of natural gas annually for thirty years. The Amu Darya deposits, with prospective reserves of 1700 bcm, were chosen as the resource base of the pipeline, and in November 2006, Turkmenistan invited the participation of CNPC in development of the Gunorta Yoloten deposit.[10] The pipeline stretches 2,500 miles from Turkmenistan's northeastern border, through Uzbekistan and Kazakhstan, and into China. China provided a $25 million low-interest loan to Turkmenistan in US dollars for purchasing Chinese drilling equipment.[11] The pipeline built on existing infrastructure, including a parallel oil pipeline from Kazakhstan to China, completed in 2005. China also suggested an additional pipeline that would reach Kazakhstan's western gas fields. The most realistic plan, however, is to create a single pipeline that would service all three Central Asian states and connect with the domestic Chinese network.[12] As of 2012, China is also pushing forward a second pipeline from Turkmenistan through Tajikistan, which will also benefit northern Afghanistan.[13]

Preconstruction estimates for construction of the pipeline from Turkmenistan to China ran as high as $20 to $30 billion. CNPC announced an investment of about $2.16 billion in the pipeline, which will connect to China's second west-east gas pipeline at Khorgos in Xinjiang Province.[14] CNPC has signed natural gas purchase agreements with Turkmenistan and a production-sharing agreement to develop the right bank of the Amu Darya River. CNPC and its Turkmen counterpart, the state-owned Turkmengeologiia, planned to drill twelve test wells to a depth of five thousand meters at Gunorta Yoloten, at a cost of $152 million.[15]

The deal with China benefits Turkmenistan's foreign policy goal of reducing dependence on Russia and Gazprom. With China's agreement to pay for the thirty-year supply of gas in US dollars, Turkmenistan can reduce dependence on Russian export routes and can start to demand higher prices from Russia for gas exports. Since Moscow resells Turkmen gas at two to three times the price that it paid, the deal with Beijing may seriously impact Russian foreign currency earnings and reduce the strategic value Gazprom plays in the region.[16] However, since the Turkmen portion of the pipeline, running 188 kilometers from the Malai gas field to the Uzbek border, was built by Stroitransgaz, a Russian firm, China feared that the company would stall development of the pipeline. The 1,300-kilometer Kazakh portion of the pipeline was built by Kaztransgaz.[17]

Turkmenistan is a natural target for Chinese investment as the second-largest gas producer and gas exporter in the former Soviet Union, after Russia. The country's gas production runs about 47 to 67 bcm per year, and is thirteenth in the world in proven gas reserves, amounting to about 2.67 tcm, although authorities claim that potential reserves are nearly ten times higher. Geological study of Turkmenistan is not highly advanced, so it is highly probable that new reserves will be found in addition to the current 139 gas deposits on the mainland and 10 on the Caspian shelf.[18] Natural gas profits contribute substantially to overall GDP, but export revenues are also redistributed to other sectors, stimulating economic growth.

As of 2009, before the opening of the pipeline to China, Turkmen exports amount to about 35 to 52 bcm per year, most of which was bound for post-Soviet markets and a small amount to Iran. Although the agricultural sector accounts for most of the country's employment, the gas and oil sectors dominate the economy. Natural gas became dominant in the 1970s, with gas output reaching 70.5 bcm per year (16.2 percent of total Soviet production) by 1980, peaking in 1989 at 89.9 bcm per year. Under the Soviet system, Turkmenistan provided natural gas to the Soviet network of pipelines, and the gas was resold to other Soviet republics at subsidized rates, depriving the country of the profits it may have earned.

At independence, Turkmenistan was the fourth-largest gas producer in the world, but political and economic turmoil led to severe cuts in the gas sector and drove down GDP.[19] Deterioration due to neglect, underinvestment, and insufficient protection from corrosion by sour gas means that the country's transmission grid is in need of significant renovation.[20] However, Turkmen domestic demand is modest: only 14 bcm per year in 2006, and 19 bcm per year predicted by 2020. Even with the uncertainty about the total size of Turkmen gas reserves, including an audit commissioned by President Niyazov that was never released to the public despite pressure from Gazprom, the proven reserves available for export are still substantial.[21]

Turkmenistan's geographical position, and the pipeline network inherited from the Soviet Union, profoundly influences the country's gas trade. The country has no export outlets to foreign markets of its own, with the partial exception of the internationally sanctioned regime in Iran, and until the new pipeline it has had to rely on Uzbekistan, Kazakhstan, and especially Russia for transit. The pipeline to Iran, which transports about 8 bcm per year, is too insignificant in relation to total gas production to reduce Turkmen dependence on the former Soviet transport system and markets. In addition, Turkmenistan is dependent for transit and export on gas-producing competitors, which further boxes in its ability to pursue an independent, profit-maximizing gas policy.[22] Turkmenistan's main gas-producing area is in the southeast region of the country, with the largest single gas reserve being Turkmengaz-owned Dauletabad, which accounts for most of the country's production and provides gas exports to Iran. Dauletabad produced about 78 million cubic meters per day in 2006. In 2003, a giant new gas field, Gunorta Yoloten, was discovered in eastern Turkmenistan, and in 2006 the nearby Osman Yoloten was discovered, and the two fields may be a single supergiant field. Turkmenistan's second, and much smaller, onshore gas-producing area is in the far west of the country, which produces about 2.7 million cubic meters per day. The Turkmen government is also encouraging investment in the Caspian offshore deposits, with about thirty-two blocks available for licensing. Rich gas fields also exist in a portion of the Caspian Sea whose ownership is disputed with Azerbaiijan.[23]

Like Turkmenistan, Kazakhstan bears the strong imprint of Soviet rule even while its resource base increasingly interests China. Soviet administrators viewed Kazakhstan, which lacked its own economically viable gas resources, purely as a transit corridor for Central Asian gas. The pipeline system was designed accordingly, failing to cover the whole of Kazakh territory and not linked into a single complex. The Kazakh Soviet Republic, unlike the Turkmen and Uzbek ones, did not control the pipelines within its borders. As a consequence of this historical development and Kazakhstan's own technical limitations, the country is now entirely dependent on Russia for access to major gas markets and its domestic gas infrastructure is inadequate.[24] The geography of Kazakhstan and the location of the gas fields also complicate the country's potential as a natural gas exporter. The gas reserves are substantial, and certainly exceed the country's expected domestic demand, which is estimated to reach about 20 to 30 bcm per year by 2020. The major gas fields are poorly located for a planned pipeline eastward to China, but well positioned for export across the Caspian Sea to Europe, being only a few hundred kilometers from existing pipelines through other former Soviet republics.[25]

According to official Kazakh figures, domestic hydrocarbon reserves contain about 3.3 to 3.7 tcm of gas, of which 2.5 tcm are proven. The country's total gas potential could reach 6 to 8 tcm with further development of resources in the Caspian Sea, which may well exceed Turkmenistan's supply. Kazakhstan's gas sector has achieved considerable production growth, from 4.34 bcm in 1994 to probably over 34 bcm in 2008. The Ministry of Energy and Mineral Resources predicts production around 80 bcm by 2015 and 114 bcm by 2020.[26] Due to the relatively sour quality of the gas from a number of Kazakh fields, only about 45 percent of the total gas volume produced in Kazakhstan is saleable, with the remainder left over after reinjection, flaring, and shipment for further purification.[27] Kazakhstan's gas reserves are almost all (97 percent) located in four western provinces, which are the home to the country's largest hydrocarbon fields and a number of smaller ones. All of these fields are close to the existing pipeline infrastructure on the traditional Russian route. Indeed, five hydrocarbon fields, all in the western provinces, contain 79 percent of Kazakhstan's total proven natural gas potential.[28]

Perhaps less ideal for China is the geography of Kazakh deposits: about 40 percent of Kazakh gas reserves are at Karachaganak, near the country's northern border with Russia. Virtually all of Karachaganak's production is processed and marketed across the border in the Russian town of Orenburg, to which it is connected by pipeline. In addition, the country has significant and underdeveloped deposits in the southeast near the border with Kyrgyzstan. The remainder of the country's gas deposits are associated with the main oil deposits. Kazakhstan continues to import gas, however, since there are no gas pipelines connecting the western gas deposits with the consumer areas in the north and east. The southern region imports gas from Uzbekistan, and the northern region from Russia.[29]

The Karachaganak gas field, the cornerstone of Kazakh gas production, was discovered in 1979 and began production in 1984. The gas was fairly deep and included significant amounts of impurities, which required treatment even before transport to refineries. In 1992, the Kazakh government established the state enterprise Kazakhmunaigas, which was responsible for exploration, production, and transport of natural gas. The government sought foreign investment early on in the Karachaganak gas field, but Gazprom was excluded from the tender, withdrew from the project, and prevented shipment of gas through its pipelines starting in the mid-1990s. The consequence was severely delayed production at the site, and Kazakhstan's gas production only increased after that time due to the growth in oil production. Gazprom sold its shares in Karachaganak to Lukoil, the Russian state-owned oil company, which invested in $5.5 billion in pipelines to connect the oil field to existing Russian pipelines. The oil field also suffered from the 1998 financial crisis in Russia, which caused production to drop by about two-thirds as customers failed to pay for gas. Work continued on a production plant and an associated pipeline, which were completed several months early in 2003.[30]

Unlike Turkmenistan and Kazakhstan, Uzbekistan is unusual among large gas producers: until very recently, the lion's share of its production was consumed domestically and only a small fraction of it was exported. Between 2000 and 2007, only 16 percent of Uzbek natural gas production was exported, compared to 73 percent for Turkmenistan and 32 percent for Russia. Potential for Uzbekistan to become a major supplier of natural gas to world markets is limited because of geological, technical, and infrastructural obstacles, as well as high domestic demand.[31] Domestic demand is already around 40 bcm per year and could reach as high as 75 bcm per year by 2020, which would not allow the country to export reserves in a volume that would make a pipeline profitable. By 2007, the country had already committed almost all of its exportable gas surplus to Kyrgyzstan, Tajikistan, or Russia.[32] Uzbekistan has a much larger population and economy than Turkmenistan does, ensuring that the country will export a smaller proportion of its natural gas production. Uzbekistan also has a much lower GDP per capita, with only about a quarter of Kazakhstan's and half of Turkmenistan's GDP.[33] By 2020, Uzbekistan plans to decrease the share of gas in its domestic fuel mix from 88 percent in 2005 to 77 percent by 2020, in part by partially replacing natural gas with coal to increase the amount of gas available for export.[34]

Importantly for China, however, Uzbekistan's gas reserves are not negligible, and the country also possesses an additional resource, like Georgia in the South Caucasus: transit potential. In 2007, Uzbekistan, with a gas output of 58.5 bcm, was the fifteenth largest gas producer in the world, and its share of total world production has been declining since 1999. Uzbekistan's proven gas reserves are an estimated 1.87 tcm, nineteenth in the world. Oil and gas operations in Uzbekistan are not separated from each other. The monopoly operator is the majority state-owned holding company Uzbekneftegaz, which was established in 1998. While Uzbekistan has allowed limited foreign participation in gas exploration projects, Uzbekneftegaz still accounts for 95 percent of total gas production.[35] As of 2005, Uzbekistan's potential gas reserves were estimated at 5.9 tcm. In total, 193 hydrocarbon fields have been discovered, of which 93 are under exploration, 67 prepared for exploration, and 33 undergoing prospecting and infrastructure development. Recoverable reserves are thinly spread over a large number of small, and occasionally tiny, deposits. Just three deposits account for about 75 percent of total gas production, but even the largest deposits are relatively small. In addition, Uzbek gas is of relatively poor quality, and about 90 percent of gas produced contains hydrogen sulfide, which can cause rapid corrosion of pipelines. Production consequently has high processing costs. The largest functioning deposits have already started to decline in production and at least one second-tier deposit has been depleted. Far more than Turkmenistan and Kazakhstan, Uzbekistan's gas production is limited by geological and technological challenges, which is a large part of the reason why the country is able to produce (at least at present) a comparatively high amount of gas but relatively little trade.[36]

SIMILAR ENERGY SECTORS

The China–Central Asia natural gas pipeline shares important commonalities with the oil and gas pipelines through Azerbaijan, Georgia, and Turkey. Like those three, especially Azerbaijan, the countries of Central Asia have large state sectors, including state monopolies (or near monopolies) over the energy sectors in each country, and consequently the energy sectors are not immune from political pressure and even arbitrary government action. As a corollary, in Central Asia and in Azerbaijan, Georgia, and Turkey, the energy sectors are not fully legal-rational. They are not entirely governed by objective regulations in which private energy actors have a significant role, and as a result, they do not fully lend themselves to harmonization on a functional basis since they are still driven, in part, by personality.

The power of the national energy sector may well be another similarity between Azerbaijan, Turkey, and the countries of Central Asia. In each country, the state oil company and state energy sector generally played driving roles not only in pipeline development but also in exploration, production, and transit of gas and oil more generally. The state oil companies own controlling shares in the largest projects and have close relationships with the branches of government that redistribute oil revenues elsewhere in government, making them enormously powerful players.

Although Azerbaijan and Georgia have made more progress than Central Asia, the former Soviet states lack a fully legal-rational governance structure of their energy sectors that can capitalize on international investment in lieu of the state energy monopoly. Turkey's state oil, gas, and pipeline companies dominate its energy sector.[37] In Azerbaijan, Georgia, and Turkey, this increasingly rational regulatory regime is cultivated and encouraged by TERAs such as EU programs and international financial institutions. As has been examined in previous chapters, TERAs are highly involved in the development of projects such as the BTC and BTE pipelines. However, apart from the involvement of TERAs, the energy sectors of Turkmenistan, Kazakhstan and Uzbekistan are strikingly similar to those of Azerbaijan, Georgia, and Turkey.

The Turkmen gas sector is a state monopoly, exercised through the state-owned oil and gas companies Turkmenneft and Turkmengaz, the Ministry of Oil and Mineral Resources, and the state agency on foreign investment. In reality, however, all crucial decisions are made by the president. In August 2008, a new law on hydrocarbon resources came into force, which formalizes direct presidential control over the gas and oil sector. The law created a state agency on the management of hydrocarbon resources under the president himself, and the agency received exclusive prerogatives in formulating energy policy and dealing with foreign investors, including conducting tenders on licenses and signing contracts.[38] Again, this seems to differ from Azerbaijan in particular only in a matter of degree.

Like Turkmenistan, Kazakhstan's oil and gas governance regime is not fully legal-rational, and foreign investors have had their share of difficulties with local and national authorities. The country's general prosecutor charged that Agip, the Italian oil company, and British Gas had underpaid taxes by about 5 billion Kazakh tenge after an audit revealed that $178 million had allegedly been concealed. The matter was resolved only with the intervention of Prime Minister Balgimbayev. In 2001, the Karachaganak consortium became embroiled in a tax dispute in which both Kazakhstan and Russia demanded value-added tax payments, which shut down production until the two governments agreed on a resolution. More recently, the consortium has been pressured to increase the number of Kazakh companies involved in gas development.[39]

Similarly to Azerbaijan, Georgia, and Turkey, Central Asian states have unresolved territorial disputes, wrangling over Caspian Sea access, or contentious issues of environmental hazards, terrorism, or the drug trade, despite their cooperation in some large-scale energy projects. Environmental concerns have also been raised, with claims of periodic release of airborne contaminants. Kazakh gas is known for its sourness, and the main fields have particularly high levels of poisonous hydrogen sulfide and other sulfates.[40] Soil tests from around Karachaganak indicate contamination with hydrogen sulfide, which reduces grain yields in the region and, according to a local university, leaves much of the soil in the area deprived of micronutrients. Two human settlements need to be moved due to the pollution.[41] The one glaring difference between the energy sectors of Azerbaijan, Georgia, and Turkey and those of the three Central Asian states discussed in this chapter is the involvement of TERAs in the development of pipeline projects in the former. The Chinese government and CNPC are not TERAs and, as has been discussed, seem to have done little to promote integration among Kazakhstan, Turkmenistan, and Uzbekistan.

A LACK OF INTEGRATION AMONG THE CHINA–CENTRAL ASIA PIPELINE COUNTRIES

The natural gas pipeline through Turkmenistan, Uzbekistan, Kazakhstan, and China has not resulted in the same level of energy and transport integration as in Azerbaijan, Georgia, and Turkey, and it is not clear that it will. As has been discussed in previous chapters, the building of the BTC and BTE pipelines contributed to the integration of the energy and transport sectors of Azerbaijan, Georgia, and Turkey. The building of the China–Central Asia gas pipeline, at the broadest level, has not resulted in similar integration, even in those limited sectors. Although, like all large transnational projects, the Central Asian gas pipeline placed some pressure on the member countries toward considering the effects of infrastructural linkages to neighboring countries, the project's development was conducted on a bilateral basis between China and each transit country separately.[42] In general, there are three underlying reasons why the integrative process does not appear to be occurring in Central Asia, despite the presence of an emerging power, China, which could potentially act as a hegemonic integrator.[43] TERAs, or rather their lack of involvement, play a role in all three areas.

First, decision-making authority in the Central Asian energy sector is highly personalized and only weakly governed by legal-rational rules. This is, in part, a function of the weak foothold that private foreign investment has in the natural gas sectors of each country, and the correlated weakness in contractual protections and objective regulations for natural gas producers. Certainly, Azerbaijan's oil sector is arguably dominated by personalized decision making outside ordinary parliamentary governance, but the involvement of TERAs limits the ability of the state to act arbitrarily.[44] Not so in Central Asia. Personalized control and the absence of objective regulatory authority do not lend the regimes to standardization across national boundaries and do not lead to integration. Harmonization of trade, customs, and administrative policies requires a transparent regulatory regime that is enforced against all actors and provides business operations with predictable legal environments and mechanisms for dispute resolution.

Second, Turkmenistan, Uzbekistan, and Kazakhstan are competitors in the global energy marketplace. Each of them produces significant volumes of natural gas, and each seeks the most profitable export routes. They do not have the natural compatibility that Azerbaijan, Georgia, and Turkey do, and they are only cooperating because, at the moment, it is profitable to do so with a sponsor in China. This does not provide a strong basis on which to integrate energy or transport sectors to the exclusion of the surrounding countries because it does not fundamentally change expectations about state behavior. The three countries vary in foreign policy orientation, particularly in their relationships with Russia, and they are not to the point where they may fully "trust" one another's motives.[45] TERAs have not been involved in the development of energy sector projects, such as the China–Central Asia pipeline, in such a way that might contribute to the building of trust, even in this limited sector. More sustained and frequent contacts among the three countries, to the exclusion of their neighbors, will be required to see real integration take place. As Chung writes, "The economic benefits of trading with China seem to have so far accrued more favorably to China than to the Central Asian countries, considering the differential terms of trade between China's export of consumer durables, foodstuffs, fertilizers, electric appliances, and machines to the region, and its import of mineral, agricultural, and animal products from there."[46]

Third, as both a cause and a consequence of the first two observations, TERAs such as supranational agencies and international financial institutions have been largely absent from the rapidly expanding natural gas sector in Central Asia and played no part in the development of the China–Central Asia pipeline. This may be due in part to the shadowy, extraregulatory nature of decision making in the natural gas sector, and it may be due in part to the fact that China and Russia are crowding out these multilateral institutions. China and Russia may also fill some of the roles played by TERAs, such as providing funding and technical assistance for large-scale projects, without the strings attached in terms of harmonization and professionalization of standards. In fact, it seems clear that China succeeded in building the China–Central Asia pipeline because it did not blend energy ventures with support for fundamental change in the modes of operation of regional energy sectors.[47]

These three explanations reflect the three integration theories outlined in chapter 1. The absence of an objective regulatory regime precludes integration from a neofunctional standpoint. Likewise, the failure to develop common expectations about state behavior contradicts a security communities theory or constructivist explanation for integration. Finally, the absence of multilateral institutions in the building of the natural gas pipeline network in Central Asia reflects neoliberal institutionalist theory. Unlike the other two theories, however, neoliberal institutionalism has an explanation for all three of these observations. In Azerbaijan, Georgia, and Turkey, integration of regulatory regimes is often cultivated or directed by supranational agencies. Emerging integrative networks change societal expectations and eventually change identities. But international institutions are reluctant to go into countries that are credit risks, in which decision making is highly personalized and opportunities for redress in the event of expropriation, corruption, or contractual violations are limited. Integration breeds integration, but the incapacity for integration is also self-sustaining.

As Keohane explained in his theory of neoliberal institutionalism, supranational institutions reduce transaction costs of further cooperation and allow the spread of information more easily. Supranational organizations establish rules for the distribution of gains, monitor the behavior of other states, and mutually sanction noncompliance. Because of the absence of supranational institutions in natural gas pipeline development in Central Asia, as well as the non–mutually reinforcing goals of the participants in the pipeline project, the risk of defection is greater than it would be in Azerbaijan, Georgia, and Turkey. That is, the cooperation of the Central Asian states, cultivated by China, does not spread benefits or reduce transaction costs equally among all of the participating states, and as a consequence, it does not necessarily guarantee sustained and repeated future cooperation. If a worthwhile alternative appears—rehabilitation of Russia-bound Soviet-era pipelines, for instance—the three countries participating in the pipeline project in Central Asia could turn away from China. The absence of TERAs in the Central Asian pipeline network contributes to a failure to institutionalize and regularize the contacts and transactions that have taken place among Turkmenistan, Uzbekistan, and Kazakhstan.

According to neoliberal institutionalism, states form collective arrangements to better share information, regularize contacts and communications, and eventually to alter state behavior and enforce norms. The institutions may take on a life of their own and cultivate further integration, institutionalizing reciprocity and making defection more costly. Institutions help states reach more mutually beneficial goals because they allow for closer, more predictable relationships among states and can set expectations accordingly. The existence of international institutions in financing and technical assistance in Azerbaijan, Georgia, and Turkey has helped to curb arbitrary government interventions and to encourage regulatory reform, encouraging integration of the three countries' energy sectors. The absence of international institutions in the development of the China–Central Asia pipeline has stunted opportunities for technical development and regulatory harmonization.

Unlike Azerbaijan, Georgia, and Turkey, Central Asian countries have not benefited as much from international technical assistance in natural gas production. Rivalries between Kazakhstan and Russia, for instance, have dampened private-sector interest in the region. In 2000, Kazakh authorities established Kaztransgaz, a gas subsidiary of the national oil transportation monopoly Kaztransoil, which took over operations of gas pipelines from a Belgian firm in a bid to regain control over the gas infrastructure. Because of periodic cutbacks in supply due to nonpayment by Russian and other customers during the financial crisis, the Belgian firm was unwilling to subsidize Kazakh gas without any governmental reforms.[48] Another consequence of the growth in production of natural gas at Kazakhstan's oil fields has been the government's increasing pressure on oil companies to use the gas productively, which may have benefited from coordination and technical assistance. Associated gas constitutes more than 45 percent of the gas produced in Kazakhstan. Aside from Karachaganak, natural gas is produced at the country's major oil fields, Tengiz, Uzen, Zhanazhol, and Kumkol. In 1999, the government passed a law requiring the inclusion of natural gas utilization projects in oil field development plans. Projects so far undertaken have included an electric power plant at Kumkol, a gas reinjection scheme at Tengiz, and a gas processing facility at Uzen. Production was only 13 bcm per year in 2002, but the government hoped to reach 47 bcm per year by 2010 and 57 bcm per year by 2015.[49]

The absence of international institutions has also failed to assist countries like Turkmenistan develop sound business regimes that invite foreign development and build creditworthy reputations. Although the Turkmen government has sought to provide a legal regime that would attract investment in oil and gas, the government provides no real protection for investors. Investment laws passed shortly after independence at least in theory protect foreign investments from nationalization or requisition, grant concessions for between five and forty years, and adhere to principles of nondiscrimination between foreign-owned firms and national ones. However, hydrocarbon resources are also national property and managed by the cabinet. Although a presidential decree provided for a body that would create rules and regulations to govern the development of hydrocarbon fields, in reality the president himself makes these key decisions.[50]

Regulatory inefficiencies prevent any real harmonization without information sharing and regular interaction among Central Asian nations (and the lack of objective legal-rational governance of the energy sectors). One of the risks faced by international gas companies is double taxation, since although production-sharing agreements or joint-venture agreements with the government of Turkmenistan control for taxation, exported gas can face a value-added tax on exports. Foreign investors are guaranteed the right of international arbitration, but national legislation curtails these rights. None of the legal frameworks governing oil and gas exploration explicitly provide for enforcement of international arbitration awards, nor do they permit a waiver of sovereign immunity in the event of asset seizure or forced transfer. The government of Turkmenistan has a history of capricious and arbitrary expropriation of property, and a number of projects (such as the trans-Afghan pipeline projects) have faced actual or threatened expropriation in the past.[51]

Finally, the lack of institutionalization in the Central Asia region reduces transparency and information sharing and engenders resistance from segments of the population. Uzbekistan suffers from the lack of public input into its energy development. Some local commentators believe that Uzbek authorities, in order to increase the supply of gas for export, are decreasing deliveries to the residential sector, which has sparked a media interest in gas supply shortages. If such a policy is being pursued, it could endanger the political and economic foundations of the present regime. The Uzbek economy is neopatrimonial, as it is based on an informal deal between the political elite and urban populations, resting on the redistribution of agricultural goods and natural resources in favor of urban populations to the detriment of rural residents. Authorities cannot cut off supplies of cheap energy and implement unlimited price increases without endangering political stability.[52]

LESS FOREIGN INVESTMENT

Foreign investment in Kazakhstan, Uzbekistan, and Turkmenistan is far more limited than in Azerbaijan, Georgia, and Turkey, which means that natural gas sectors are driven by state-owned companies that do not necessarily operate under objective and legal-rational regulation. State-owned oil and gas companies are understandably reluctant to share information with their counterparts in other countries, a gap that TERAs may help redress. This is not to say, however, that they lack any foreign investment at all. Like oil, the development of natural gas in Kazakhstan has been constrained by transport difficulties and the vagaries of Russian-Kazakh relations. Development of the gas sector will depend on foreign investment, including both those investing in the natural gas fields and those investing in oil reserves that also contain gas deposits.[53] International oil companies entered Kazakhstan on privileged terms at a time of low oil prices when Kazakhstan did not have a developed oil or gas sector, unlike Russia or Azerbaijan. The country signed its first agreement with a foreign company in 1993, a joint venture with Chevron in the Tengiz oil field. Unlike any other former Soviet gas producer, with the partial exception of Azerbaijan, international oil companies play a central role in the gas sector in Kazakhstan. In 2007, two international consortia were responsible for more than 70 percent of Kazakhstan's gas output. In 2002, President Nazarbaev signed a decree establishing the national oil and gas company Kazmunaigaz, which became a vehicle for channeling the government's influence into an energy sector dominated by foreign companies. In 2004, legislation was introduced requiring that Kazmunaigaz have a minimum 50 percent share in new PSAs.[54] The Kazakh government needs international oil and gas companies for their technical expertise in developing the geologically sophisticated Caspian hydrocarbon fields, but the government does have a strong say in influencing the shape and structure of export routes. Although Kazmunaigaz is only responsible for about 6 to 8 percent of domestic gas output, it controls the key export gas pipelines through the subsidiary Kaztransgaz.[55]

The lack of foreign investment, and the strong involvement of the Russian energy sector, is also evident in Uzbekistan. In 2007, only an estimated 5 percent of Uzbekistan's total gas output was produced with participation by foreign capital. So far, the largest foreign investors are Russian and offshore companies, although since the mid-2000s, Asian companies are increasing their presence. The combination of an insular government, a large state sector, and the belief in Uzbekistan that the oil and gas sector is important for security reasons led to a preference for state self-reliance at the expense of foreign investment in energy.[56] In 2006, the Uzbek government loosened its grip on the oil and gas sector and allowed foreign participation in the privatization of strategic enterprises. The largest foreign investors in Uzbekistan's gas sector are Russian companies, particularly Gazprom and Lukoil. Lukoil is the largest foreign investor in Uzbek gas projects, including hundreds of millions of dollars in investments during the 2000s, and a controlling share in a gas field that will begin production in 2013. Because Gazprom monopolizes the Russian gas pipeline network, Lukoil in particular has been expanding its stake in gas production and export in former Soviet republics like Azerbaijan, Kazakhstan, and Uzbekistan in order to resell its gas to Gazprom at higher prices. The key investment roles played by Gazprom and Lukoil belie the overall lack of foreign direct investment in the energy sectors of Kazakhstan, Turkmenistan, and Uzbekistan. These are in many ways "legacy" investors, due to links from the Soviet past, and generally not reliant on the role of TERAs. Outside investment from elsewhere would be much more dependent on the "stamp of approval" that TERAs provide.

INTERNATIONAL ORGANIZATIONS BUT LACK OF INTEGRATION

While there is little evidence of integration among Kazakhstan, Turkmenistan, and Uzbekistan's energy sectors, in the context of this counterfactual, it is worth examining the various attempts at regional integration in Central Asia, at least one of which purports to be an energy-oriented grouping. It is also worth looking at these attempts because they all take the form of international organizations that are similar, at least in their outward appearances, to TERAs. It is worth noting from the outset, however, that Turkmenistan, due to its avowed neutrality, is not nor has it ever been a member of any of these international organizations. While this immediately negates the potential role of these organizations in encouraging integration among Turkmenistan, Kazakhstan, and Uzbekistan, it is still worth looking at attempts at integration in the region, to examine whether they have had any effect on the energy sectors of the three countries.

The China-led Shanghai Cooperation Organization (SCO) is probably the most prominent of these institutional attempts to bring integration to the region. Members include Russia as well as the four post-Soviet Central Asia states, excluding Turkmenistan. While its official objectives do not mention "integration" as such, it brings members together for regular military exercises and coordinates counterterrorism activities through a center in Tashkent. In 2006, then–Russian President Vladimir Putin suggested that the SCO form an "energy club" due to the perceived complementarity of Russian and Central Asian oil and gas resources with China's demand. By all accounts, this idea was not received favorably by other SCO members, mainly due to their desire to keep their energy sectors as independent as possible from Russian influence.[57] The Russia-led Collective Security Treaty Organization (CSTO) is, as its name suggests, almost exclusively focused on military and security issues. Of the three Central Asian states in question, it includes Kazakhstan but not Turkmenistan, and Uzbekistan left and reentered the organization in the 1990s and 2000s. The CSTO primarily serves as an umbrella for a number of bilateral security agreements between Russia and the other members, some of which allow Russian forces on their soil. Some also participate in regular military exercises and high-level summits, although Uzbekistan has notably not been a regular participant in these activities. The treaty itself does not mention energy and although the Russian national security strategy considers energy and security issues to be closely linked, there have not been major discussions about energy sector integration under CSTO auspices.[58] One might expect the Russia-led Eurasian Economic Community (EAEC) to encourage linkages among energy sectors in Central Asia, but in general it is concerned with the trade of consumer goods and equipment. One of its stated aims is to form a common energy market and it has within its institutions structures such as an energy council. That said, of the Central Asia countries discussed here, only Kazakhstan is an active member; Turkmenistan has never been a member and Uzbekistan's membership is "suspended." Uzbekistan only joined in 2006 and its leadership quickly became disillusioned with the project. EAEC summits have discussed the possibility of merging with the CSTO, which seems to support the general analysis that the EAEC is mostly a vehicle for Russia to maintain influence with Central Asian leaders.[59]

As Roy Allison has termed it, these organizations provide a "protective integration," tied to either China or Russia, prioritizing regime security, sovereignty, and resistance to what is perceived by Central Asian leaders as "external values." Allison argues that "bandwagoning" with China or Russia "has spelled the end of efforts by Central Asian states at regional self-organization."[60] Indeed, it would seem that compared to the role of TERAs in Azerbaijan, Georgia, and Turkey, which previous chapters have shown encourage integration among the three countries, even if only in discrete sectors, the international organizations of Central Asia do exactly the opposite. Moreover, the China-led SCO played no discernible role in developing the China–Central Asia pipeline.[61]

The Chinese government also played a limited role, even as a hegemonic integrator. China's negotiations with the three countries involved were not only bilateral, but the terms of the agreements were not disclosed to the other two pipeline transit states. Construction of the three segments of the pipeline was undertaken as three separate projects, with separate Chinese-led consortia responsible. The only time that the leaders of Kazakhstan, Turkmenistan, and Uzbekistan met in relation to the China–Central Asia pipeline was when the project had finished construction and was officially opened for operation. Apart from connecting the segments of the pipeline at border crossings (some of which already existed due to Soviet-era links), there is no evidence to suggest that the China–Central Asia pipeline's construction plans played any role in integrating the energy sectors of Kazakhstan, Turkmenistan, and Uzbekistan. The project did not go forward under the auspices of or with the support of any TERAs. As discussed above, the TERAs active in Azerbaijan, Georgia, and Turkey—TRACECA, INOGATE, and the World Bank—are also active in Central Asia, but none of them were involved in the China–Central Asia pipeline. Other TERAs also active in the region, such as the Asian Development Bank, did not play a role in the pipeline's completion.[62]

CONCLUSION

Azerbaijan, Georgia, and Turkey are developing an increasingly intertwined energy sector in which technical requirements are standardized and regulatory instruments are compatible. This is not occurring among Kazakhstan, Turkmenistan, and Uzbekistan, where the natural gas pipeline network, in which China has heavily invested for its own energy needs, is governed by a highly personalized and state-centric structure. Turkmenistan, Uzbekistan, and Kazakhstan lack the natural compatibility of Azerbaijan, Georgia, and Turkey and are engaging in an arms-length pipeline project that is not changing the nature of their energy sectors on the ground. This is in large part because of the absence of TERAs such as EU programs or the World Bank in the development of the China–Central Asia pipeline. Although China is an extra-regional power that is "encouraging" the development of the Central Asian pipeline network, it is doing so in a way that favors its energy balance. It does not share the same priorities as TERAs active in Azerbaijan, Georgia, and Turkey and it has interacted with the three Central Asian states on a bilateral basis.

Other contributing factors include the highly personalized decision-making authority in the energy sectors of Kazakhstan, Turkmenistan, and Uzbekistan, only weakly governed by legal-rational rules. This absence of an objective regulatory regime precludes integration from a neofunctional standpoint. TERAs are not involved to play a role in limiting highly personalized, arbitrary decision making. The three countries are also competitors in the global energy marketplace. Their foreign energy policies could change rapidly and opportunistically, just as they did when they joined the China–Central Asia pipeline project. This failure to develop common expectations about state behavior goes against a major security communities theory explanation for integration. TERAs have not been involved in the development of the China–Central Asia pipeline in such a way that might contribute to the building of trust, even in this limited sector. Also, relative to Azerbaijan, Georgia, and Turkey, TERAs have been much less involved in energy sector development in Central Asia overall, not just with the China–Central Asia pipeline. This relative absence of multilateral institutions reflects the key component identified by neoliberal institutionalist integration theory.

Despite the integrationist rhetoric of hegemon-driven international institutions in Central Asia, such as the SCO and the CSTO, Kazakhstan, Turkmenistan, and Uzbekistan do not all take part in these organizations and they have had little discernable impact on the development of the China–Central Asia pipeline. TERAs active in Central Asia have not been involved in the pipeline's development either. This leads to the conclusion that while there are a number of important contributing factors as to why the three Central Asia states have not integrated their energy sectors the way that Azerbaijan, Georgia, and Turkey have, the absence of TERAs in the development of major pipeline projects such as the China–Central Asia pipeline is an important explanatory variable. One can further conclude that the role of TERAs in Azerbaijan, Georgia, and Turkey is also an important explanatory variable as to why integration has occurred in the energy and transport sectors, particularly as the role of TERAs fits into the three theories of integration employed in this study.

NOTES

1. "China National Petroleum Subsidiaries to Pay Billions for Central Asia Gas Pipeline," *China Post*, December 30, 2007.

2. Simon Pirani, "Turkmenistan: An Exporter in Transition," in *Russia and CIS Gas Markets and Their Impact on Europe*, ed. Simon Pirani (Oxford: Oxford University Press, 2009), 295.

3. Kathleen Hancock, "Escaping Russia, Looking to China: Turkmenistan Pins Hopes on China's Thirst for Natural Gas," *China and Eurasia Forum Quarterly* 4, no. 3 (2006): 78–79.

4. Thrassy N. Marketos, *China's Energy Geopolitics: The Shanghai Cooperation Organization and Central Asia* (New York: Routledge, 2009), 15–16.

5. See, e.g., Niklas Swanstrom, "China and Central Asia: A New Great Game or Traditional Vassal Relations?," *Contemporary China* 14, no. 45 (2005): 569–84, and Janet Xuanli Liao, "A Silk Road for Oil: Sino-Kazakh Energy Diplomacy," *Brown Journal of World Affairs* 7, no. 2 (2006): 39.
6. Shamil Yenikeyeff, "Kazakhstan's Gas Sector," in Pirani, *Russia and CIS Gas Markets*, 345–47.
7. Marketos, *China's Energy Geopolitics*, 17.
8. Ibid., 16.
9. Stanislav Zhukov, "Uzbekistan: A Domestically Oriented Gas Producer," in Pirani, *Russia and CIS Gas Markets*, 389.
10. Pirani, "Turkmenistan," 295–96.
11. Hancock, "Escaping Russia, Looking to China," 77.
12. Jonathan Stern and Michael Bradshaw, "Russian and Central Asian Gas Supply for Asia," in *Natural Gas in Asia: The Challenges of Growth in China, India, Japan, and Korea*, ed. Jonathan Stern, 2nd ed. (Oxford: Oxford University Press, 2008), 261–62.
13. Marketos, *China's Energy Geopolitics*, 17, and Vladimir Socor, "Beijing Proposes Turkmenistan-China Gas Pipeline through Northern Afghanistan," *Eurasia Daily Monitor*, June 19, 2012.
14. Pirani, "Turkmenistan," 296.
15. Ibid., 295–96.
16. Marketos, *China's Energy Geopolitics*, 120.
17. Pirani, "Turkmenistan," 296–98.
18. Ibid., 271.
19. Ibid., 274.
20. Ibid., 284.
21. Stern and Bradshaw, "Russian and Central Asian Gas Supply," 260–61.
22. Pirani, "Turkmenistan," 288.
23. Ibid., 279–282.
24. Yenikeyeff, "Kazakhstan's Gas Sector," 316.
25. Stern and Bradshaw, "Russian and Central Asian Gas Supply," 261.
26. Yenikeyeff, "Kazakhstan's Gas Sector," 320–21.
27. Ibid., 322.
28. Ibid., 324.
29. Anne E. Peck, *Economic Development in Kazakhstan: The Role of Large Enterprises and Foreign Investment* (New York: Routledge, 2004), 193–94.
30. Ibid., 194–98.
31. Zhukov, "Uzbekistan," 355.
32. Stern and Bradshaw, "Russian and Central Asian Gas Supply," 261.
33. Pirani, "Turkmenistan," 271, 274.
34. Zhukov, "Uzbekistan," 386.
35. Ibid., 357, 359.
36. Ibid., 360–61.
37. "Turkey: Energy Sector Review," INOGATE, www.inogate.org.
38. Pirani, "Turkmenistan," 284.
39. Eric Watkins, "Karachaganak Oil Now Subject to Export Tax," *Oil and Gas Journal*, July 23, 2011.
40. Yenikeyeff, "Kazakhstan's Gas Sector," 322.
41. Peck, *Economic Development in Kazakhstan*, 197–98.
42. Marketos, *China's Energy Geopolitics*, 120.
43. As has been noted above and by Keohane, this would primarily result in integration with the hegemon as opposed to with one another. This does not seem to be the case with TERAs in Azerbaijan, Georgia, and Turkey. The integration is occurring among the three countries.
44. For instance, the presidents of both countries control the oil funds of Azerbaijan (SOFAZ, the State Oil Fund of the Azerbaijani Republic) and Kazakhstan (NFRK, the National Fund for the Republic of Kazakhstan), which redistribute oil profits to the

state to other developmental sectors. Yelena Kalyuzhnova, "Overcoming the Curse of Hydrocarbon: Goals and Governance in the Oil Funds of Kazakhstan and Azerbaijan," *Comparative Economic Studies* 48, no. 3 (2006): 600.

45. Edward Chow (senior fellow, Energy and National Security Program, Center for Strategic and International Studies), interview with the author, July 5, 2007.

46. Chien-peng Chung, "The Shanghai Co-operation Organization: China's Changing Influence in Central Asia," *China Quarterly* 180 (2004): 1002.

47. Andrew E. Kramer, "New Gas Pipeline from Central Asia Feeds China," *New York Times*, December 15, 2009.

48. Yenikeyeff, "Kazakhstan's Gas Sector," 318.

49. Peck, *Economic Development in Kazakhstan*, 198.

50. Martha Brill Olcott, "International Gas Trade in Central Asia: Turkmenistan, Iran, Russia, and Afghanistan," in *Natural Gas and Geopolitics: From 1970 to 2040*, ed. David G. Victor, Amy M. Jaffe, and Mark H. Hayes (Cambridge: Cambridge University Press, 2006), 210.

51. Olcott, "International Gas Trade in Central Asia," 210–11.

52. Zhukov, "Uzbekistan," 387.

53. Peck, *Economic Development in Kazakhstan*, 195.

54. Yenikeyeff, "Kazakhstan's Gas Sector," 318–19.

55. Ibid., 319.

56. Zhukov, "Uzbekistan," 373.

57. Sergei Kolchin, "Russia Initiates SCO Energy Club," RIA Novosti, June 21, 2006.

58. Roy Allison, "Virtual Regionalism, Regional Structures and Regime Security in Central Asia," *Central Asian Survey* 27, no. 2 (2008): 193–95.

59. Ibid., 192–93.

60. Ibid., 198.

61. Daniel Kimmage, "Turkmenistan-China Pipeline Project Has Far-Reaching Implications," *Radio Free Europe/Radio Liberty*, April 10, 2006.

62. See, from the Asian Development Bank, Harinder S. Kohli, Ashok Sharma, and Anil Snood, eds., *Asia 2050: Realizing the Asian Century* (New Dehli: SAGE, 2011).

Conclusion

Azerbaijan, Georgia, and Turkey are currently engaged in a process of integration in their energy and transport sectors, helping to harmonize regulatory regimes across borders, to develop common strategies of growth and development, and to improve the business climate by ensuring predictability and standardization across the three countries. The integration currently underway in the energy and transport sectors has not placed pressure on other sectors to follow suit, the way that coal and steel integration in Western Europe in the 1950s reduced trade barriers and opened the door to broad economic integration across the continent.[1] This may be because of the unique characteristics of these sectors. First, energy and transport in the three countries are highly regulated industries that possess strong state presence, even monopoly or near-monopoly control, in which there are a finite number of actors and the means of ensuring high rates of compliance. Second, given the political economy of the three countries and the geography of transporting oil and gas to world markets, actors in the energy and transport sectors in Azerbaijan, Georgia, and Turkey have strong financial incentives to lower transaction costs and produce rational, transparent regimes. Third, Azerbaijan, Georgia, and Turkey possess a natural complementarity in which each possesses a comparative advantage as to the other two countries: Azerbaijan is well endowed with a high-value natural resource; Georgia has a transportation infrastructure that is necessary to bypass the closed borders with Armenia and Iran and avoid Russian export routes; and Turkey has deep, sustained access to world (especially European) markets and international investors. These unique characteristics are helping to integrate Azerbaijan, Georgia, and Turkey in a distinctive and limited way.

Integration involves (a) at least rudimentary institutional structures among the countries involved; (b) at least some interdependence in policy formation and implementation; and (c) at least some collective sense of mutual identity and obligation.[2] The more depth that the integration process achieves, the greater the penalties will become for defection, the lower the costs of cooperation, and the more predictable the state's behavior will be to outside actors. More precisely, as used in this study, integration is the process by which a group of nation-states become interdependent as to one another to the exclusion of those not in the group. Through legal and regulatory harmonization, political and social intertwinement, and the erection of institutions common to the three countries, the members of the group will become more peaceful as to one another and will reduce barriers to economic development. Azerbaijan, Georgia, and Turkey, to the exclusion of bordering neighbors such as Russia, Iran, and Armenia, are currently drawn into a reinforcing cycle of integration that is changing expectations about predictability, reducing economic barriers, and inviting foreign investment.[3]

The three traditional theories of integration that originally developed to explain the progressive unification of Western European in the decades following the Second World War attempted to fill a gap in traditional realist theory that had difficulty explaining why sustained and durable integration was replacing balance of power notions and shifting alliances. As coal and steel harmonization pressured other regulatory sectors toward harmonization, an enormously complex supranational regime developed that nearly spanned the continent. As the supranational institutions grew, they developed their own center of gravity and cultivated further integration. However, outside the European context, this pattern of integration was not repeated in the same way. The Southern African Development Community, for instance, has not generated self-sustaining momentum toward unification, and the Association of Southeast Asian Nations still remains a highly state-centric organization that generally abides by principles of noninterference among member states.[4] The inability to precisely reproduce the theoretical integration of Western Europe in other regions of the world led to continued refinements and layers to the traditional theories of integration.

The first of these theories, security communities theory, was originally conceived by Karl Deutsch in the influential book *Political Community and the North Atlantic Area*, first published in 1957. Deutsch first observed clusters of states with relatively peaceful relationships despite the reigning balance of power, and he sought to isolate the variable making this possible. According to Deutsch, international relations were not solely limited to interactions among states, but as a world of overlapping communities, peoples, and identity groupings, making it an early forerunner to the constructivist revolt against realist theory following the Cold War.[5] He proposed that transactions and communications among different communities, when considered in the aggregate, could eventually build levels of trust and interdependence that would change collective identity and create a "we-feeling" among people across borders. He referred to these emerging groups of countries as "security communities," and successive scholarship defined in more precise terms the different stages or types of security communities.[6] For instance, although it is possible that emerging security communities could be warlike as to other security communities, some scholars have shown that the more tightly knit and integrated a security community, the more the community projects shared norms and common interests externally so that security communities ultimately produce more peaceful outcomes. While Deutsch's original theory contained methodological limitations, the essential core has remained unchanged: namely, that a state's behavior is learned and could change as the shared norms of the community change. Security communities theory is generally able to import advances in sociology and human behavior into international relations theory, which helps provide additional variables to measuring the progress of integration.

The second traditional theory of integration is neofunctionalism, which posits that regulatory harmonization at low technocratic levels will place pressure on higher political levels and other regulatory spheres to progressively continue the integrative process (a process called "spillover"). The theory used mechanisms similar to the prewar theory of functionalism as conceived by David Mitrany, which sought to create international technical institutions that would expand their power, engineer a shift in loyalties from nation-states to the international institutions, and eventually form a peaceful community of interdependent nations.[7] Neofunctionalism, however, was conceived as a regional theory rather than a global one, and is considerably more complex, according to the original theorists Ernst B. Haas and Leon Lindberg, who studied the integration of the coal and steel sectors in Western Europe and analyzed how that integration spilled over into energy integration through Euratom and economic integration through the European Economic Community. The theory suffered a setback in the early 1960s as French President Charles de Gaulle succeeded in paralyzing the governing commission of the EEC and thereby slowing the process of European integration, an episode that underscored that nonfunctional politics could undermine the regulatory harmonization process and revealed that the integrative process was not unidirectional and deterministic.[8] In addition, spillover does not seem to have occurred outside the European context, at least not as much. Although neofunctionalist theory underwent several revisions, it survives in the present more as a hybrid theory or "lens" rather than a comprehensive explanation for integration. The value of neofunctionalism is that it identifies actors and processes that are not identified in other theories, such as the role of supranational elites, business organizations and trade unions, and regulatory harmonization.

The third traditional theory of integration, and the one most relevant to this analysis, is neoliberal institutionalism as conceived by Robert Keohane in his 1984 book *After Hegemony*. According to Keohane, nation-states set up supranational institutions in order to achieve common goals, but, once built, these institutions could alter state behavior by reducing transaction costs, establishing and enforcing norms, sharing information, and regularizing contact and communications.[9] The institutions could cultivate their own growth and dominance and engineer integration from the top down. Participation in such institutions can help a state build its reputation or creditworthiness among other members and can increase the costs of defection and reduce the costs of compliance. Rooted in realist presumptions that states are the primary actors in international relations and will act in their own interest, neoliberal institutionalism departs from realism by positing that supranational institutions change state behavior over time by altering norms and expectations and consequently state interests. The theory introduces a new mechanism for integration, with supranational institutions as the variable, although, like neofunctionalism, it recognizes that supranational elites often cultivate and encourage further integration in a self-reinforcing cycle. As used in this study, transnational extra-regional actors (TERAs) are a specific kind of supranational institution: namely, those from outside the region that pursue broad-based agendas beyond simple profit and influence for the benefit of states that are not core members of the institutions.[10]

The integration currently occurring among Azerbaijan, Georgia, and Turkey is limited in nature, restricted to the high-value and heavily regulated spheres of energy and transport. Highly regulated sectors lend themselves to standardization across borders, and the integrative process increases the pressure for governments to ensure transparency of finances and to respect contracts with private partners and resist expropriation. The organic integration that is occurring among these three countries is cultivated as a top-down process by TERAs operating with broad-based agendas, which provide technical assistance and financial support to integrative projects. However, this integrative process has not led to integration in other sectors. Politically, for instance, Azerbaijan, Georgia, and Turkey do not yet share common attitudes toward neighbors such as Armenia and Russia, nor do they share a common geopolitical orientation toward Europe. On a cultural level, while two of the three countries share a common Turkic language (Azerbaijan and Turkey), two are secular and Islamic (Azerbaijan and Turkey), two are oriented toward the European Union (Georgia and Turkey), and two share a common history of Russian and Soviet domination (Azerbaijan and Georgia), the three countries together have very few common cultural threads as distinguished from their neighbors. On a security level, the countries do not yet possess complementary goals, as evidenced by the 2008 war between Georgia and Russia during which Turkey was unable or unwilling to come to Georgia's defense.[11] Thus far, integration has been limited to technical realms and has not generated momentum toward broader integration in a manner similar to that of Western Europe.

The process of integration is not necessarily consistent or unidirectional, and even in the energy and transport sectors the integrative process has not progressed in the same manner. The energy sector, particularly in Azerbaijan, is of a shadowy nature and largely outside the reach of parliament. TERAs that wish to influence development of the energy sector must do so at the very highest levels of the executive branch, because it is the executive branch that negotiates concession agreements, divides and distributes state oil revenues, and makes important decisions about Azerbaijan's foreign energy relations.[12] By contrast, the transport sector is decentralized, accessible to the ordinary population, intertwined with other actors such as border control and customs, and dependent on local cooperation.[13] TERAs operating in the transport and communications sector have multiple avenues and pressure points on which they can bring their influence to bear. Because of the different makeup of these two sectors, the integrative processes in each are best explained by slightly different theories. Undoubtedly, neoliberal institutionalism best explains integration in the energy sector, where public perception and expectations are much less relevant because of the closed nature of the sector. For the transport sector, however, a hybrid theory is necessary. Neoliberal institutionalism accounts for the top-down nature of the integrative process, but public perceptions and expectations play some role as in security communities theory. In addition, because the sector is accessible and decentralized, regulatory harmonization has concrete results that affect a great number of actors. While this is still a centrally planned process of integration, encouraged by international institutions, the mechanisms identified by neofunctionalism and security communities theory are not absent.

The study analyzed two TERAs that are contributing to integration among Azerbaijan, Georgia, and Turkey: first, the European Union–sponsored projects Transport Corridor Europe-Caucasus-Asia (TRACECA) and Interstate Oil and Gas Transport to Europe (INOGATE), and second, the World Bank's efforts at promoting harmonization and transparency in the energy sector. In order to isolate TERAs as the variable promoting integration, the study proposed a counterfactual involving the natural gas pipeline network from Turkmenistan to China. Like the oil and natural gas pipelines from Baku to Ceyhan and Erzerum, the natural gas pipeline was an extraordinary feat of engineering designed to link an immense reservoir of natural resources with world markets. The Central Asian project, however, was driven almost solely by China's desire to cheaply close its energy gap. The project did not benefit from technical or financial assistance from TERAs. The result is that very little integration has occurred among Kazakhstan, Turkmenistan, and Uzbekistan—the transit countries—where a lack of transparency and the overwhelming control of the state energy company ensure that energy is treated as sui generis to a greater degree than even Azerbaijan, outside normal political processes.[14] As a result, the pipeline is unlikely to have any effect on integration in the energy sector or other sectors and is unlikely to spur cooperation among the countries in realms unrelated to the pipeline.

IMPLICATIONS AND SHORTCOMINGS

The explanatory variable used in this analysis, namely, the presence of TERAs in cultivating and encouraging integration from the top down, emphasizes a strain of integration theory that holds that integration is not an organic, progressive, and deterministic process without orchestration from above. In other words, the integrative process is not spontaneous and continuous; rather, it is directed by the agency of individuals and institutions.[15] The thesis is that integration has had a profound impact on the business climate for foreign energy companies, which has helped spur foreign direct investment and thereby promote economic development, but may have contributed to an overemphasis on energy development to the detriment of other sectors of the economy and society.[16] Integration of the energy sector in particular has helped solidify predictability of the government's range of action and has reduced the potential for expropriation, as neoliberal institutionalism would predict: integration changes state behavior. One way in which the proposed railway stretching from Baku to Tbilisi and then to Kars is helping produce positive consequences for the region is that it is revitalizing northeastern Anatolia, a long-neglected part of Turkey. The railway may be considered a "crucial lifeline" according to local officials.[17] Although some scholars have termed this process "geographical spillover," the expansion of economic development due to large-scale infrastructure construction is not solely a product of neofunctionalist theory.

One methodological limitation with using transnational extra-regional actors as a variable is that the distinctive quality of TERAs (namely, that they are extra-regional) is not essential to the variable itself. In other words, an international institution based in the same region as the countries under study could potentially be capable of encouraging integration in the same way as TERAs, provided that it had similar access to technical assistance and donor funding or credit. Similarly, although the TERAs under study here, namely projects of the European Union and the World Bank, possess broad-based agendas that aim to improve governance and transparency in client countries, a broad-based agenda is likewise not essential to the definition of TERAs or to the process of integration. That said, a multinational corporation or a hegemonic neighboring state does not act similarly to international institutions in general or to TERAs in particular, because they do not cultivate integrative processes beyond transparency or security for their own resource-extraction and profit goals, with the exception of occasional philanthropy. To this end, TERAs fit comfortably in the theory of neoliberal institutionalism because they fit Keohane's model of international institutions as state-created organizations that possess an institutional agency and a legal personhood separate from their member states.

A second methodological limitation of the thesis is that it only identifies a strong correlation between the presence of international institutions and integration among Azerbaijan, Georgia, and Turkey, and identifies a mechanism by which international institutions appear to be feeding the integrative process. The study does not necessarily prove, or attempt to prove, causation, which is not possible at present because of the short span of time covered by the study. The counterfactual case of the natural gas pipeline from Turkmenistan to China possesses its own methodological limitations in attempting to isolate TERAs as an explanatory variable. It is not, however, the only variable. According to Transparency International's 2011 corruption index, for instance, Turkmenistan (177th out of 182 countries) and Uzbekistan (tied for 177th) rank far worse than Georgia (64th) or Turkey (62nd) and slightly worse than Azerbaijan (143rd).[18] It is possible that the lack of structural institutional transparency has prevented TERAs from operating in the region, so it becomes difficult to prove that the *absence* of TERAs *contributes to* the lack of governmental transparency. However, this study contends that TERAs are still an explanatory variable, as there is a strong correlation between their presence and integration in the sectors in which they operate. One benefit of the "hybrid" nature of the theory used in the present study is that it incorporated a broad array of actors that were not reached exclusively by a single theory. As an empirical matter, the study looked at communications and transactions among populations, at the role of business interests and supranational elites, and at external relations (e.g., common aversion to Russia, a neighboring regional hegemon) before settling on the role of international institutions as the explanatory variable.

SUMMING UP

Integration in the energy and transport sectors among Azerbaijan, Georgia, and Turkey is an organic process that is changing the relationships among the three countries, even if only in a limited way. It is not, however, a spontaneous process, nor an inevitable one. Rather, it is cultivated from the top down by international institutions that are providing technical assistance and financial support to large-scale, cross-border projects that are changing expectations and promoting a more predictable business climate. Azerbaijan possesses significant reserves of highly valued natural resources; Georgia serves as a geographically crucial transport corridor; and Turkey has deep and sustained access to world markets and international partners. Integration among the three countries reinforces their existing complementarity. However, outside the energy and transport sectors, the three countries have not made significant progress toward integration. The three countries do not share similar regional orientations, linguistic or religious backgrounds, or even common foreign relations toward neighbors such as Russia, Armenia, and Iran. Most importantly, integration in the energy and transport sectors has not placed pressure on other sectors to integrate as well; rather these two sectors have been treated as distinctive from others precisely because they have integrated, in a self-reinforcing but limited process.

The explanatory variable for integration among the three countries is the presence of transnational extra-regional actors, which include European Union and World Bank projects designed to improve transparency and governance in the energy sector and promote transport and freight capacity through the Caucasus and Turkey. The role of TERAs in promoting and cultivating integration in the transport and energy sectors accords with neoliberal institutional theory, which posits that states join international institutions because it is in their national interest to do so, but over time their presence in the organization will start to shape state behavior by changing the risks and rewards of membership and increasing interactions among the members. The role of TERAs is particularly notable in the energy sector, which has historically been a closed sector, accessible only to the highest levels of government. By contrast, integration in the transport sector involves a much broader range of actors, as the sector is decentralized and accessible to the ordinary population. Consequently, a theory that not only incorporates neoliberal institutionalism but also elements of security communities theory, namely sociological interactions among the general population, as well as neofunctionalism in the form of regulatory standardization, presents additional variables that help explain multifaceted integration in the transport sector. The three countries under study here—Azerbaijan, Georgia, and Turkey—are forming a limited bond that unites the three of them to the exclusion of their neighbors. While integration is not necessarily a unidirectional or progressive process, the complementarity of the three countries, as well as incentives to promote integration, undoubtedly have an impact on the tempo and depth of the integrative process currently underway. Only time will tell whether the results are irreversible or transferable to other sectors or regions, but, although this pattern of integration thus far is distinctive to these three countries, the range of variables identified in this study undoubtedly have relevance to other contexts.

NOTES

1. Desmond Dinan, *Ever Closer Union: An Introduction to European Integration* (Boulder, CO: Lynne Rienner, 2005), 26–38.

2. Joseph S. Nye, "Multinational Enterprises and Prospects for Regional and Global Political Integration," *Annals of the American Academy of Political and Social Science* 403 (1973): 117–19.

3. See Slavica Penev, "Investment Climate and Foreign Direct Investment Trends in the South Caucasus and Central Asia," *South East European Journal of Economics and Business* 2, no. 1 (2007): 34–35 (explaining impact of pipelines and transit projects on economic and business climate in the region).

4. See, e.g., Richard Higgott, "Alternative Models of Regional Cooperation? The Limits of Regional Institutionalization in East Asia," in *European Union and New Regionalism: Regional Actors and Global Governance in a Post-Hegemonic Era*, ed. Mario Telò (Burlington, VT: Ashgate, 2007), 75–106; Fredrik Soderbaum, "African Regionalism and EU-African Interregionalism," in Telò, *European Union and New Regionalism*, 185–202; Kjell A. Eliassen and Catherine Børve Arnesen, "Comparison of European and South Asian Integration," in Telò, *European Union and New Regionalism*, 203–221.

5. Karl W. Deutsch, *Political Community and the North Atlantic Area: International Organization in the Light of Historical Experience* (Princeton, NJ: Princeton University Press, 1957).

6. See generally, Emanuel Adler and Michael Barnett, eds., *Security Communities* (New York: Cambridge University Press, 1998).

7. Neofunctionalism combines the mechanism of functionalism with federalist goals. See Jeppe Tranholm-Mikkelsen, "Neo-functionalism: Obstinate or Obsolete? A Reappraisal in Light of the New Dynamism of the EC," *Millennium* 20, no. 1 (1991): 1–22.

8. Laura Cram, "Integration Theory and the Study of the European Policy Process: Towards a Synthesis of Approaches," in *European Union: Power and Policy Making*, ed. Jeremy Richardson, 2nd ed. (New York: Routledge, 2001), 59–60. Later scholarship has even questioned the importance of the European Coal and Steel Community (ECSC) to the development of the European Economic Community given that the ECSC's legacy was quite mixed even as to the coal and steel sectors. John Gillingham, "The European Coal and Steel Community: An Object Lesson?," in *Europe's Post War Recovery*, ed. Barry Eichengreen (New York: Cambridge University Press, 1995), 166–67.

9. Robert O. Keohane, *After Hegemony: Cooperation and Discord in the World Political Economy* (Princeton, NJ: Princeton University Press, 2005).

10. For an example of the expansive agendas of international financial institutions, see Devesh Kapur, "Expansive Agendas and Weak Instruments: Governance Related Conditionalities of International Financial Institutions," *Journal of Economic Policy Reform* 4, no. 3 (2001): 207–41.

11. Ziya Onis and Suhnaz Yilmaz, "Between Europeanization and Euro-Asianism: Foreign Policy Activism in Turkey during the AKP Era," *Turkish Studies* 10, no. 1 (2009): 7–24.

12. See, e.g., Alec Rasizade, "Azerbaijan after Heydar Aliev," *Nationalities Papers* 32, no. 1 (2004): 137–64 (generally discussing lack of transparency in energy sector).

13. Samuel Lussac, "The Baku-Tbilisi-Kars Railroad and Its Geopolitical Implications for the South Caucasus," *Caucasian Review of International Affairs* 2, no. 4 (2008): 212–24.

14. Alexander Cooley, "Principles in the Pipeline: Managing Transatlantic Values and Interests in Central Asia," *International Affairs* 84, no. 6 (2008): 1171–88.

15. This was recognized even in the European literature, with some scholars, critical of neofunctionalism (or at least a certain caricature of neofunctionalism), pointing toward the actions of the Council Presidency in encouraging or holding back European integration. Human agency plays a prominent role in the integration process. Emil Joseph Kirchner, *Decision-making in the European Community: The Council Presidency and European Integration* (New York: Manchester University Press, 1992).

16. Richard M. Auty, "Optimistic and Pessimistic Energy Rent Deployment Scenarios for Azerbaijan and Kazakhstan," in *Energy, Wealth and Governance in the Caucasus Central Asia: Lessons Not Learned*, ed. Richard Auty and Indra do Soysa (New York: Routledge, 2006), 57–76.

17. Mehmet Dikkaya and Deniz Özyakişir, "Development Regional Cooperation Among Turkey, Georgia, and Azerbaijan: Importance of Regional Projects," *Perceptions* 13, nos. 1–2 (2008): 105.

18. Transparency International, "Corruption Perceptions Index 2011," http://cpi.transparency.org/cpi2011/results/.

Bibliography

Interviews

Akhundov, Fuad (Sector Director, Social and Political Affairs Department, President's Office, Republic of Azerbaijan), 24 May 2006, Baku.
Aliriza, Bulent (Director and Senior Associate, Turkey Project, Center for Strategic and International Studies), 3 July 2007, Washington DC.
Broad, Oliver (Risk Analyst, BP Azerbaijan), 22 May 2006, Baku.
Broad, Oliver (Risk Analyst, BP Azerbaijan) and Elchin Hagvardiyev (Risk Analyst, BP Azerbaijan), 26 June 2007, Baku (by telephone).
Cholst, Anthony (Lead Country Officer, South Caucasus Country Management Unit, Europe and Central Asia Region, the World Bank), 10 August 2007, Washington DC.
Chow, Edward C. (Senior Fellow, Energy Program, Center for Strategic and International Studies), 5 July 2007, Washington DC.
Cornell, Svante (Research Director, Central Asia Caucasus Institute, Johns Hopkins University), 4 February 2009, Washington DC.
Gogia, Giorgi (Caucasus Analyst, International Crisis Group), 26 May 2006, Tbilisi.
Goltz, Thomas (author, *Azerbaijan Diary*, *Georgia Diary*; Adjunct Professor, Central and Southwest Asia Studies Program, University of Montana), 1 April 2007, London (by telephone).
Hadjy-zadeh, Hikmet (Vice President, FAR Center), 3 February 2009, Baku (by telephone).
Ismail, Anar (Analyst, TRACECA), 5 February 2009, Baku (by telephone).
Jervalidze, Liana (Chair of the Board, Georgia for Transit Development, Transparency and Regional Integration [GDTRI]), 29 May 2006, Tbilisi.
Johnson, Larry (Security Manager, BP Georgia), 28 May 2006, Tbilisi.
Kakchekeev, Bakyt (Country Director-Kyrgyzstan, Task Force Innovations International), 2 February 2009, Bishkek (by telephone).
Karakaya, M. Zahid (Advisor, BOTAS International Ltd), 13 April 2007, Ankara.
Kundakci, Utku (Assistant Program Officer, Foreign Policy Program, Turkish Economic and Social Studies Foundation {TESEV]), 17 April 2007, Istanbul.
Nutter, C. J. (Chief, Defense Threat Reduction Office, Georgia), 29 May 2006, Tbilisi.
Ozel, Soli (Professor of International Relations, Istanbul Bilgi University), 17 April 2007, Istanbul.
Roberts, John (Energy Security Specialist, Platts), 10 April 2009, London (by telephone).
Rondeli, Alexander (President, Georgian Foundation for Strategic and International Studies [GFSIS]), 26 May 2006, Tbilisi.
Smith, Gare A. (Partner, Foley Hoag LLP), 12 July 2007, Washington DC.
Starr, S. Frederick (Chairman of the Central Asia-Caucasus Institute, Johns Hopkins University), 10 February 2009, Washington DC.
Thomson, Peter (Director, Department for Sustainable Development, Europe and Central Asia Region, the World Bank), 10 August 2007, Washington DC.
Ziyadov, Taleh (Deputy Executive Director, US-Azerbaijan Chamber of Commerce), 17 August 2007, Washington DC.

Documents

"12th Athens Conference: Conclusions" (2008), Energy Community, Vienna, European Union, 16 May.
"2006 Brussels Conference: Towards an EU External Energy Policy" (2006), European Commission, Brussels, European Union, 20 November.
"Address by H. E. Ahmet Davutoglu, Minister of Foreign Affairs of Republic of Turkey at the OSCE Ministerial Council Meeting, 1–2 December 2009, Athens" (2009), *Turkish Government News*, 6 July, available at http://infoweb.newsbank.com/resources/doc/nb/news/13125419D157C608?p=AWNB.
"Albania 2008 Progress Report 2008. [SEC(2008) 2692]" (2008), Commission of the European Communities, Brussels, http://ec.europa.eu/enlargement/pdf/press_corner/key-documents/reports_nov_2008/albania_progress_report_en.pdf, last accessed 25 March 2009.
"Armenian expert downplays Turkey's role in Caucasus and Central Asia" (2001), *Noyan Tapan News Agency*, 5 October, available at http://w3.nexis.com/new/results/docview/docview.do?docLinkInd=true&risb=21_T15042308927&format=GNBFI&sort=BOOLEAN&startDocNo=26&resultsUrlKey=29_T15042308931&cisb=22_T15042308930&treeMax=true&treeWidth=0&csi=10962&docNo=44.
"As Russia moves in, EU bickers over fees" (2008), *International Herald Tribune*, 25 January.
"Azerbaijan Country Analysis Brief" (2007), Energy Information Agency, http://www.eia.doe.gov/cabs/Azerbaijan/pdf.pdf, last accessed 3 March 2009.
"Azerbaijan Develops Transport Infrastructure under TRACECA" (2008), Trend—Daily News, 4 December.
"Azerbaijan Emerging as Exporter of Gas" (2010), *Xalq Qazeti*, 13 October, available at http://w3.nexis.com/new/results/docview/docview.do?docLinkInd=true&risb=21_T15050124890&format=GNBFI&sort=BOOLEAN&startDocNo=1&resultsUrlKey=29_T15050124894&cisb=22_T15050124893&treeMax=true&treeWidth=0&csi=10962&docNo=1.
"Azerbaijan Highway Project: Project Initiation Document" (2001), *Azeravtoyol*, 27 March.
"Azerbaijan to Launch Telecoms Satellite by 2011" (2008), *Azerbaijan Press Agency*, November 12. http://w3.nexis.com/new/search/homesubmitForm.do
"Azeri Daily Criticizes Turkey's Handling of Georgian-Russian Conflict" (2008), *Zerkalo*, 21 August, http://w3.nexis.com/new/results/docview/docview.do?docLinkInd=true&risb=21_T15051200060&format=GNBFI&sort=BOOLEAN&startDocNo=1&resultsUrlKey=29_T15051200066&cisb=22_T15051200064&treeMax=true&treeWidth=0&csi=10962&docNo=17.
"Azeri Official Believes Europe-Caucasus-Asia Corridor is Victim of Indifference" (2000), BBC Monitoring—Central Asia, London, 22 September.
"Azeri surprise at Iran's protest over Caspian oil exploration projects" (2001), *Turan News Agency*, 23 July, http://w3.nexis.com/new/results/docview/docview.do?docLinkInd=true&risb=21_T15042682491&format=GNBFI&sort=BOOLEAN&startDocNo=1&resultsUrlKey=29_T15042682495&cisb=22_T15042682494&treeMax=true&treeWidth=0&csi=10962&docNo=24.
"Azeri Think-Tank Group Looks into Projects Aimed at Stability in the South Caucasus" (2008), *Turan News Agency*, 14 October, http://w3.nexis.com/new/results/docview/docview.do?docLinkInd=true&risb=21_T15048257067&format=GNBFI&sort=BOOLEAN&startDocNo=1&resultsUrlKey=29_T15048257071&cisb=22_T15048257070&treeMax=true&treeWidth=0&csi=10962&docNo=1.
"The Baku Initiative" (2008), European Commission, available from http://ec.europa.eu/dgs/energy_transport/international/regional/caspian/energy_en.htm, accessed 19 March 2009.
"Bosnia and Herzegovina 2008 Progress Report 2008. [SEC(2008) 2693 final]" (2008), Commision of the European Communities, http://ec.europa.eu/enlargement/pdf/

press_corner/key-documents/reports_nov_2008/bosnia_herzegovina_progress_report_en.pdf , last accessed 25 March 2009.

"Brief on Key Conclusions from the 4th Bosphorus Conference: 'EU and Turkey: Drifting Apart?'" British Council.

"BTC Pipeline: Oil Window to the West" (2005), Central Asia-Caucasus Institute & Silk Road Studies Program, http://www.silkroadstudies.org/new/inside/publications/BTC.pdf, last accessed 3 April 2009.

"Caspian Basin: EU Funds Environmental Protection Program for Caspian Basin" (2009), EurasiaNet, 19 December, available at http://www.eurasianet.org/departments/insightb/articles/eav121908c.shtml, accessed 16 March 2009.

"Caspian Pipeline Consortium Brochure," http://www.cpc.ru/portal/alias!press/lang!en-US/tabID!3357/DesktopDefault.aspx, 3 April 2009.

"Central Asian Regional Economic Cooperation," ADB Homepage, http://www.adb.org/Carec/transportation.asp, last accessed 16 April 2009.

"Conference in Azeri Capital Looks into Ways of Accession to WTO" (2010), *Zerkalo*, 28 October, http://w3.nexis.com/new/results/docview/docview.do?docLinkInd=true&risb=21_T15049900408&format=GNBFI&sort=BOOLEAN&startDocNo=1&resultsUrlKey=29_T15049900412&cisb=22_T15049900411&treeMax=true&treeWidth=0&csi=10962&docNo=1.

"Construction of Turkish Part of BTK Railway Set to Begin in July" (2008), *Turkish Weekly*, 15 May, http://www.turkishweekly.net/news/55328/construction-of-turkish-part-of-btk-railway-set-to-begin-in-july.html, accessed on 3 April 2009.

"Country Report: Azerbaijan" (2005), Commission Staff Working Paper "European Neighbourhood Policy" (Brussels: Commission of the European Communities), 10, http://ec.europa.eu/world/enp/pdf/country/azerbaijan_country_report_2005_en.pdf.

"Country Report: Georgia" (2005), Commission Staff Working Paper "European Neighbourhood Policy" (Brussels: Commission of the European Communities), 12, http://ec.europa.eu/world/enp/pdf/country/georgia_country_report_2005_en.pdf.

"Croatia 2008 Progress Report 2008. [SEC(2008)2694]" (2008), Commission of the European Communities, 11 May, available at http://ec.europa.eu/enlargement/pdf/press_corner/key-documents/reports_nov_2008/croatia_progress_report_en.pdf, last accessed 25 March 2009.

"Declaration of Intent for the Establishment of Competitive Regional Electricity Market in South Eastern Europe" (1999), Ministerial Conference, Thessalonki, 10 September, http://www.kepa.uoa.gr/Thessaloniki_Declaration_1999.pdf, last accessed 23 March 2009.

"Delegation from Central Asia visits the Secretariat" (2008), 23 September, http://www.energy-community.org/portal/page/portal/ENC_HOME/NEWS/News_Details?p_new_id=1261, last accessed 25 March 2009.

"Delegation from Central Asia visits the Secretariat" (2009), 23 September, http://www.energy-community.org/portal/page/portal/ENC_HOME/NEWS/News_Details?p_new_id=1261, last accessed 19 March, 2009.

"Edison, Depa Set Up Company for Italy-Greece Gas Pipeline" (2008), *Downstream Today*, 11 July.

"Energy Affairs" (2008), Greek Ministry of Foreign Affairs, December, http://www.mfa.gr/www.mfa.gr/en-US/Economic+Diplomacy/Energy+Affairs/, last accessed 30 March 2009.

"Energy Community Discusses Enlargement and Implementation" (2008), *Europe Energy*, 9 July.

"Energy Community—Institutions" (2009), *Energy Community*, 25 March, http://www.energy-community.org/portal/page/portal/ENC_HOME/ENERGY_COMMUNITY/Institutions, last accessed 25 March 2009.

"Energy Community Investment Conference, Athens" (2007), Background Information Package, 28 September.

"The Energy Community Treaty (Council Decision 2006/500/EC of 29 May 2006 on the conclusion by the European Community of the Energy Community Treaty)" (2006), European Union, 29 May.
"Energy in South East Europe—Regional Policy for a Regional Market" (2006), Energy in South East Europe, 23 November, http://www.seerecon.org/infrastructure/sectors/energy/index.html, last accessed 24 March 2009.
"Energy Policy Needs Social Dimensions" (2008), European Federation of Public Service Unions, Brussels, EPSU, 30 May.
"Energy Profile of Azerbaijan" (2007), *Encyclopedia of the Earth*, http://www.eoearth.org/article/Energy_profile_of_Azerbaijan, last accessed 30 March 2009.
"Energy Sector Development Strategy in the Context of Economic Development in Armenia" (2005), Yerevam Government of Armenia, 23 June.
"Energy Security and the PKK Threat to the BTC Pipeline" (2008), Jamestown Foundation, http://www.jamestown.org/programs/gta/single/?tx_ttnews[tt_news]=5170&tx_ttnews[backPid]=167&no_cache=1, last accessed 3 April 2009.
"Energy Superpower Emerges in the Caspian" (2008), *Asia Times Online*, 17 October, http://www.atimes.com/atimes/Central_Asia/JJ17Ag04.html.
"Eni Gazprom Sign Gas Pipeline Accord for EU" (2007), *Enerpub Energy Publisher*, 23 June, http://www.energypublisher.com/article.asp?id=10031, last accessed 1 April 2009.
"ENi loses Kashagan lead role, Kazakhstan ups stakes" (2008), Reuters UK, 14 January, http://uk.reuters.com/article/oilRpt/idUKL1412693520080114, last accessed 15 April 2009.
"The EU and Turkey: Drifting Apart? Key Conclusions of the 4th Bosphorus Conference" (2007), the British Council, www.cer.org.uk/pdf/conclusion_bosphorus_conference_2007.pdf, last accessed 19 March 2009.
"EU and Turkey: Together for a European Energy Policy" (2007), keynote speech by Andris Piebalgs, EU Energy Commissioner, to the conference on "Turkey and the EU," 5 June.
"An EU Energy Security and Solidarity Action Plan," European Commission, COM (2008) 744/3, http://news.bbc.co.uk/2/shared/bsp/hi/pdfs/14_11_08euenergy.pdf.
"EU Energy Commissioner Backs Ukraine's Gas Transit Aims" (2008), BBC Monitoring International, 3 February.
"EU Energy Policy and Turkey" (2007), European Commission Memo, June.
"EU Energy Security and Solidarity Action Plan: 2nd Strategic Energy Review" (2008), European Commission, Brussels, European Union, 13 November.
"EU starts negotiations on Caspian pipeline to bring gas to Europe" (2011), EU Neighbourhood Info Centre, 9 December, http://www.enpi-info.eu/maineast.php?id_type=1&id=26324&lang_id=450&utm_source=4marketing.it%2C%204dem.it&utm_medium=Email%20Marketing&utm_content=Subscriber%23165&utm_campaign=EU%20starts%20negotiations%20on%20Caspian%20pipeline%20to%20bring%20gas%20to%20Europe.
"EU Unveils Plan to Reduce Russia's Grip on Energy" (2008), *Guardian*, 13 November.
"EU, Turkey at Odds over Energy Community" (2007), *Hurriyet Turkish Daily News*, 7 June.
"EU, Turkey play up Energy to Keep Accession Lights On" (2007), *Ekathemirini*, 6 June.
"EU/NIS: Commission Approves TACIS Assistance Projects for Former USSR" (1997), *European Report*, Brussels, 27 September, vol. 27, issue 2254.
"EU/NIS: INOGATE Summit Opens Energy Umbrella Agreement over NIS" (1999), *European Report*, Brussels, 24 July.
"Europe's Underwater Chemical Dump" (2006), *Spiegel Online*, 30 August, http://www.spiegel.de/international/0,1518,434329,00.html, last accessed 15 April 2009.
"European Neighbourhood Policy: Black Sea Synergy" (2007), House of Commons Select Committee on European Scrutiny. London, Government of the United Kingdom of Great Britain and Northern Ireland, 19 April.

"EU-Turkey Relations," European Union Commission on Enlargement, European Union, http://ec.europa.eu/enlargement/candidate-countries/turkey/eu_turkey_relations_en.htm, last accessed 25 March 2009.
"Evaluation of Tacis Inter-State Energy and INOGATE Programmes and Related Actions implemented in the Framework of National Programmes – ref. 951599," available at http://ec.europa.eu/europeaid/how/evaluation/evaluation_reports/evinfo/2000/951599_ev_en.pdf, accessed 16 March 2009.
"Evaluation of TACIS Inter-State Energy and INOGATE Programmes and Related Actions Implemented in the Framework of National Programmes" (2000), TACIS, Final Report, volume 1, September.
"Evaluation of TACIS Inter-State Energy and INOGATE Programmes and Related Actions Implemented in the Framework of National Programmes" (2000), TACIS, Final Report, volume 2, September.
"Evaluation of the Tacis Interstate TRACECA Programme" (1998), Tacis, Evaluation Unit, June.
"Evaluation of the TRACECA Interstate Program" ref. 951421, accessible on the EU website at http://ec.europa.eu/europeaid/how/evaluation/evaluation_reports/evinfo/1998/951421_ev_en.pdf, last accessed 16 March 2009.
"'Fierce War' Between Armenia and Azerbaijan: Gas Supplies to Yerevan Cut" (1993), British Broadcasting Corporation, 25 January. Supplied by LexisNexis on 26 June 2012, http://w3.nexis.com/new/search/homesubmitForm.do.
"The Former Yugoslav Republic of Macedonia 2008 Progress Report 2008. [SEC(2008) 2695]" (2008), Commission of the European Communities, http://ec.europa.eu/enlargement/pdf/press_corner/key-documents/reports_nov_2008/bosnia_herzegovina_progress_report_en.pdf, last accessed 25 March 2009.
"Frequently Asked Questions," http://www.nabucco-pipeline.com/press-public-news/q-a/q-a.html, last accessed 31 March 2009.
"FSU Refineries: Turkmenistan's Refineries" (2000), *Alexander's Gas & Oil Connections*, http://www.gasandoil.com/goc/company/cnc02746.htm, last accessed 8 April 2009.
"Gas Export via Baku-Tbilisi-Erzurum Pipeline Ceased" (2008), Caspian Partnership for the Future, 13 August, http://www.cpf.az/xeber/254-gas-export-via-baku-tbilisi-erzurum.html.
"Gazprom Reveals Unaffordable Costs of South Stream" (2009), *Eurasia Daily Monitor* 6:29, http://www.jamestown.org/single/?no_cache=1&tx_ttnews[swords]=8fd5893941d69d0bc3f378576261ae3e&tx_ttnews[any_of_the_words]=South%20Stream&tx_ttnews[tt_news]=34495&tx_ttnews[backPid]=7&cHash=9422e92d79, last accessed 17 April 2009.
"Georgia–Russia Conflict Changes the Energy Equation" (2008), *EU Business*, 3 September, http://www.eubusiness.com/Energy/georgia-russia-energy.08-09-03/, last accessed 3 April 2009.
"Georgian Rail Bridge Blast Affects Azeri Oil Shipment" (2008) Azerbaijan News Service TV, trans. by BBC Monitoring International Reports, 17 August, Custom 150 Journals, 26 June 2012.
"Georgian Turkish Military Cooperation Develops" (2002) *Armenpress*, 10 June, http://asbarez.com/46977/georgian-turkish-military-cooperation-develops/.
"Georgia's Ultimate Goal is to Join Europe" (2011), Rustavi-2 TV, 7 October, http://w3.nexis.com/new/results/docview/docview.do?docLinkInd=true&risb=21_T15050805769&format=GNBFI&sort=BOOLEAN&startDocNo=1&resultsUrlKey=29_T15050805773&cisb=22_T15050805772&treeMax=true&treeWidth=0&csi=10962&docNo=24.
"Germany, Russia: The Nord Stream Pipe Dream" (2008), Stratfor, 13 November, http://www.stratfor.com/analysis/20081113_germany_russia_nord_stream_pipe_dream, last accessed 2 April 2009.
"Guler Says Turkey Has Reservations Regarding EU Energy Community," *Turkish Daily News* via Asia Africa Intelligence Wire; LexisNexis, 27 October 27.

"Independent Audit Shows Turkmen Gas Field 'World Class,'" (2008), Radio Free Europe, 14 October, http://www.rferl.org/content/Independent_Audit_Shows_Turkmen_Gas_Field_WorldClass_/1329822.html, last accessed 16 April 2009.
"Interview with Slavtcho Neykov, Director of the Energy Community Secretariat" (2008), *Europe Energy Magazine*, 8 October.
"Italy: Edison Rises with Poseidon" (2008), Stratfor, February 27. http://www.stratfor.com/analysis/italy_edison_rises_poseidon, last accessed 30 March 2009.
"Joint Conference Statement on 'Turkey and the EU: Together for a European Energy Policy,'" (2007), 5 June.
"Kars-Tbilisi Agreement" (2007), *Railway Gazette*, 1 February, http://www.railwaygazette.com/news_view/article/2007/02/7020/news_in_brief-8.html, last accessed 3 April 2009.
"Kars-Tbilisi-Baku railway construction to begin with ceremony in Tbilisi 20 Sep" (2007), *Anatolia News Agency*, 13 August, http://w3.nexis.com/new/results/docview/docview.do?docLinkInd=true&risb=21_T15042308927&format=GNBFI&sort=BOOLEAN&startDocNo=1&resultsUrlKey=29_T15042308931&cisb=22_T15042308930&treeMax=true&treeWidth=0&csi=10962&docNo=13.
"Kashagan field start up date delayed to 2014" (2008), *Oil and Gas Journal*, 11 August.
"Kashagan Partners Eye $4 bil. Trans-Caspian Transport System to Connect to BTC Pipeline" (2007) HIS Global Insight, http://www.globalinsight.com/SDA/SDADetail6096.htm last accessed 1 April 2009.
"Kazakh Minister Urges Boosting Energy Ties Between EU, Central Asia" (2007), *Interfax-Kazakhstan*, 7 April, http://w3.nexis.com/new/results/docview/docview.do?docLinkInd=true&risb=21_T15050694036&format=GNBFI&sort=BOOLEAN&startDocNo=1&resultsUrlKey=29_T15050694041&cisb=22_T15050694040&treeMax=true&treeWidth=0&csi=10962&docNo=18.
"Kazakhstan—Astana's Resource Nationalism Hitting ENI at Kashagan and Others" (2008), Allbusiness, 21 July, http://www.allbusiness.com/energy-utilities/energy-utility-sector-performance-oil/11421674-1.html, last accessed 15 April 2009.
"Kazakhstan Country Analysis Brief" (2008), Energy Information Administration, http://www.eia.doe.gov/cabs/Kazakhstan/pdf.pdf, last accessed 30 March 2009.
"Kazakhstan Drops Oil Refinery Plans in Georgia" (2008), Reuters UK, 24 September, http://uk.reuters.com/article/idUKTRE48N0Z720080924.
"Kazakhstan Urges Shanghai bloc to set Single Energy Market" (2008), ITAR-TASS, 30 October 30, online edition.
"Keeping the Door Wide Open; Turkey and EU Accession" (2008), Special Report of Session, House of Commons (UK) Business and Enterprise Committee, 7 October.
"Kosovo (UNDER UNSCR 1244/99) 2008 Progress Report. [SEC(2008) 2697 Final]" (2008), Commission of the European Communities, http://ec.europa.eu/enlargement/pdf/press_corner/key-documents/reports_nov_2008/kosovo_progress_report_en.pdf, last accessed 25 March 2009.
"Memo on the Energy Council" (2007), European Union Energy Council, Brussels: European Union, 5 June.
"Memorandum of Understanding on the Regional Electricity Market in South East Europe and its Integration into the European Union Internal Electricity Market" (2002), http://www.seerecon.org/infrastructure/documents/mou-rem-see.pdf, last accessed 24 March 2009.
"Ministerial Declaration on Enhanced Energy Co-operation between the EU, the Littoral States of the Black and Caspian Seas and their Neighbouring Countries" (2006), Baku Initiative, Astana, European Union, 30 November.
"Montenegro 2008 Progress Report 2008. [SEC(2008) 2696]" (2008), Commision of the European Communities, http://ec.europa.eu/enlargement/pdf/press_corner/key-documents/reports_nov_2008/montenegro_progress_report_en.pdf, last accessed 25 March 2009.

"More Trouble for BP as Gas Scheme is Halted" (2007), *Guardian*, 5 February, http://www.guardian.co.uk/business/2007/feb/05/oilandpetrol.news, last accessed 17 April 2009.

"MP Says Russia Ready to Negotiate with Georgian Ministers" (2010), *Rezonansi*, 8 October, http://w3.nexis.com/new/results/docview/docview.do?docLinkInd=true&risb=21_T15048463066&format=GNBFI&sort=BOOLEAN&startDocNo=1&resultsUrlKey=29_T15048463070&cisb=22_T15048463069&treeMax=true&treeWidth=0&csi=10962&docNo=1.

"Nabucco Survives as EU Priority Project" (2009), *Russia Today*, 21 March, http://www.russiatoday.com/Business/2009-03-21/Nabucco_survives_as_EU_priority_project.html, last accessed 31 March 2009.

"Nazad – v budushchee, ili v poiskakh evraziiskovo transportnogo konsensusa" (Back - to the future, or in search of the Eurasian transport consensus), http://www.ref.uz/referat.php?id=1857, accessed on 16 March 2009.

"Negotiation Mandate: Enhanced EC-Ukraine Agreement" (2007), European Parliament, Strasbourg, European Union, 12 July.

"Objectives of the Marmaray Project," Marmaray homepage, http://www.marmaray.com/html/gen_obj.html, last accessed 3 April 2009.

"Obzor TRASEKA i dostizhenia proekta garmonizatsii protsediur peresecheniia granits"(Review of TRACECA and the Achievements of the Project's Harmonization of Border Crossing Procedures) (2003), Asian Development Bank report for a Regional Seminar on Trade Facilitation and Customs Modernization, August 4–8, Issyk-Kul, Kyrgyz Republic.

"Officials Address Kashagan Field Delays" (2008), *European Weekly*, 15 September, http://www.neurope.eu/articles/89746.php, last accessed 14 April, 2009.

"Oil and Gas in Azerbaijan," http://www.azerb.com/az-oil.html, last accessed 30 March 2009.

"Oil and Gas: Turkey" (1999), *EIU Business Middle East*, 1 November, http://w3.nexis.com/new/results/docview/docview.do?docLinkInd=true&risb=21_T15040008591&format=GNBFI&sort=BOOLEAN&startDocNo=26&resultsUrlKey=29_T15040014912&cisb=22_T15040014911&treeMax=true&treeWidth=0&csi=253477&docNo=28.

"Overview of INOGATE Projects (1996–2008)" (2008) INOGATE, http://www.inogate.org/projects?collection=1996_2008&lang=en, accessed 16 March 2009.

"Partnership and Cooperation Agreement; Armenia" (1999), European Council and Commission, Brussels, European Union, 9 September.

"Partnership and Cooperation Agreement; Azerbaijan" (1999), European Council and Commission, Brussels, European Union, 5 May.

"Petroleum Geology and Resources of the North Ustyurt Basin" (2001), *U.S. Geological Survey Bulletin*, http://pubs.usgs.gov/bul/2201/D/b2201-d.pdf, last accessed 30 March 2009.

"President Barroso and the President of Azerbaijan sign a Memorandum of Understanding on Energy Partnership" (2006), European Council and Commission, Brussels, European Union, 7 November.

"Progress of White Stream" (2007), GUEU White Stream Pipeline Project, http://www.energy-community.org/pls/portal/docs/91811.PDF, last accessed 1 April 2009.

"Project Description/ Pipeline Route," Nabucco Gas Pipeline Project homepage, http://www.nabucco-pipeline.com/project/project-description-pipeline-route/project-description.html, last accessed 31 March 2009.

"Project Timeline," Nabucco Gas Pipeline Project homepage, http://www.nabucco-pipeline.com/project/project-timeline/main-page-project-timeline-20090126.html, last accessed 31 March 2009.

"Report on the Implementation of the Acquis under the Treaty Establishing the Energy Community" (2008), European Community Secretariat, May, http://www.energy-community.org/pls/portal/docs/118175.PDF, last accessed 25 March 2009.

"Report on the Implementation of the Treaty" (2007), presented at 5th Permanent High Level Group Meeting on 28 June 2007 and 2nd Ministerial Council Meeting on 29 June 2007 in Becili, Montenegro, 30 May, available at http://www.energy-community.org/pls/portal/docs/72182.PDF, last accessed 25 March 2009.

"Republic of Turkmenistan," United Nations Economic Commission for Europe, http://www.unece.org/operact/opera/sppled/tur.htm, last accessed 8 April 2009.

"Resolution of 23 October 2008" (2008), European Parliament, Strasbourg, European Union, 23 October.

"Russia Builds Second Military Base in Armenia" (2008), *Ayna*, 3 May, http://w3.nexis.com/new/results/docview/docview.do?docLinkInd=true&risb=21_T15051200060&format=GNBFI&sort=BOOLEAN&startDocNo=1&resultsUrlKey=29_T15051200066&cisb=22_T15051200064&treeMax=true&treeWidth=0&csi=10962&docNo=19.

"Russia Kazakhstan Agree on Expansion of CPC Pipeline" (2009), *Zawya*, 3 April, http://www.zawya.com/story.cfm/sidv51n19-3NC02/Russia,%20Kazakhstan%20Agree%20On%20Expansion%20Of%20CPC%20Pipeline, last accessed 3 April 2009.

"Russia: The BTC Pipeline's Surprising Savior" (2005), Stratfor, January 6, http://www.stratfor.com/br_russia_btc_pipelines_surprising_savior, accessed 27 March 2009.

"Russia's Top Law-Enforcement Officials Discuss Caucasus Drug Trafficking" (2010), *Interfax*, 25 February, http://w3.nexis.com/new/results/docview/docview.do?docLinkInd=true&risb=21_T15049824356&format=GNBFI&sort=BOOLEAN&startDocNo=1&resultsUrlKey=29_T15049824360&cisb=22_T15049824359&treeMax=true&treeWidth=0&csi=10962&docNo=3.

"Russian Experts Skeptical About Nabucco Gas Pipeline Project" (2009), *Ekho Moskvy*, 13 July, http://w3.nexis.com/new/results/docview/docview.do?docLinkInd=true&risb=21_T15050321071&format=GNBFI&sort=BOOLEAN&startDocNo=1&resultsUrlKey=29_T15050321075&cisb=22_T15050321074&treeMax=true&treeWidth=0&csi=10962&docNo=2.

"Russian Paper Expects European Security to Change After Georgia War" (2008), *Gazeta*, 26 August, http://w3.nexis.com/new/results/docview/docview.do?docLinkInd=true&risb=21_T15049367204&format=GNBFI&sort=BOOLEAN&startDocNo=1&resultsUrlKey=29_T15049367209&cisb=22_T15049367208&treeMax=true&treeWidth=0&csi=10962&docNo=23.

"Russian paper interviews Azeri deputy foreign minister on Caspian developments" (2009), *Izvestiya*, 30 November, http://w3.nexis.com/new/results/docview/docview.do?docLinkInd=true&risb=21_T15048643018&format=GNBFI&sort=BOOLEAN&startDocNo=1&resultsUrlKey=29_T15048643022&cisb=22_T15048643021&treeMax=true&treeWidth=0&csi=10962&docNo=8.

"Russian TV Portrays GUAM Summit as US-Funded Bid to Weaken CIS" (2006), *Vesti*, 23 May, http://w3.nexis.com/new/results/docview/docview.do?docLinkInd=true&risb=21_T15049969956&format=GNBFI&sort=BOOLEAN&startDocNo=1&resultsUrlKey=29_T15049969960&cisb=22_T15049969959&treeMax=true&treeWidth=0&csi=10962&docNo=7.

"SCP Commissioning Commences" (2006) BP homepage, 1 June, http://www.bp.com/genericarticle.do?categoryId=9006615&contentId=7018471, last accessed 27 March 2009.

"Secretariat Intensifies Contacts with Azerbaijan," available from: Energy Secretariat Website http://www.energy-community.org/portal/page/portal/ENC_HOME/NEWS/News_Details?p_new_id=1081 accessed 19 March 2009.

"Securing Your Energy Future: Commission Presents Energy Security, Solidarity and Efficiency Proposals" (2008), European Commission, Brussels, European Union, 13 November.

"Serbia 2008 Progress Report 2008 [SEC(2008) 2698 final]" (2008), Commission of the European Communities, http://ec.europa.eu/enlargement/pdf/press_corner/key-documents/reports_nov_2008/serbia_progress_report_en.pdf, last accessed 25 March 2009.

"Serbia Urges EU to Stop Setting New Entry Requirements" (2008), *VOA News*, 6 November, http://www.voanews.com/english/2008-11-06-voa63.cfm, last accessed 25 March 2009.

"Several ethnic Azeri activists arrested in Iran" (2008), *Turan News Agency*, 31 July, http://w3.nexis.com/new/results/docview/docview.do?docLinkInd=true&risb=21_T15042914149&format=GNBFI&sort=BOOLEAN&startDocNo=1&resultsUrlKey=29_T15042914153&cisb=22_T15042914152&treeMax=true&treeWidth=0&csi=10962&docNo=18.

"Shah-Deniz Consortium and GIOC Sign Gas Export Deals" (2003), *Alexander's Gas & Oil Connections*, 8:23, http://www.gasandoil.com/goc/company/cnc34805.htm, last accessed 30 March 2009.

"Shah Deniz taps primed" (2006), Upstream online.com, 14 September, http://www.upstreamonline.com/live/article119108.ece, last accessed 27 March 2009.

"SOCAR 2006 Annual Report" (2006), SOCOR homepage, http://socar.az/others-en.html, last accessed 3 March 2009.

"Southern Russia unsure about government's immigration policy" (2005), *Yuzhnyy Reporter*, 21 November, http://w3.nexis.com/new/results/docview/docview.do?docLinkInd=true&risb=21_T15042308927&format=GNBFI&sort=BOOLEAN&startDocNo=1&resultsUrlKey=29_T15042308931&cisb=22_T15042308930&treeMax=true&treeWidth=0&csi=10962&docNo=21.

"Status of the Nord Stream Pipeline Route in the Baltic Sea" (2007), Nord Stream Project Information, October, http://www.nord-stream.com/uploads/media/Nord_Stream_Route_Status_ENGLISH.pdf, last accessed 2 April 2009.

"Strategic Goals and Objectives" (2009), Foreign Policy Strategy 2006–2009 (Tbilisi: Ministry of Foreign Affairs of Georgia), 7–17, http://www.mfa.gov.ge/files/35_9440_673620_11.pdf.

"A Strong European Neighborhood Policy" (2007), European Commission, Brussels, European Union, 12 May.

"A Strong European Neighbourhood Policy" (2007), Communication from the Commission (Brussels: Commission of the European Communities), December 5, 2–4, http://ec.europa.eu/world/enp/pdf/com07_774_en.pdf.

"Supsa Terminal and Pipeline, Georgia/Azerbaijan," http://www.hydrocarbons-technology.com/projects/supsa/, last accessed 30 March 2009.

"Third Ministerial Council Press Release" (2007), Energy Community, Belgrade, European Union, 18 December.

"TRASEKA nachala realizatsiiu novogo proekta v stranakh SNG" (TRACECA Began Implementing a New Project in NIS) (2008), Kazinform, 20 October, http://atyrauobl.kz/traseka-nachala-realizaciyu-novogo-proekta-v-stranax-sng/2702, accessed 16 March 2009.

"The Treaty Establishing the Energy Community (ECT)" (2005), 25 October, http://www.seerecon.org/infrastructure/sectors/energy/documents/Treatyenergy.pdf, last accessed 24 March 2009.

"Turkey 2008 Progress Report 2008. [SEC(2008) 2699]" (2008), Commission of the European Communities, http://ec.europa.eu/enlargement/press_corner/key-documents/reports_nov_2008_en.htm, last accessed 25 March 2009.

"Turkey, Europe: Nabucco No Longer an Empty Pipe Dream?" (2008), Stratfor, February 5, http://www.stratfor.com/analysis/turkey_europe_nabucco_no_longer_empty_pipe_dream, last accessed 31 March 2009.

"Turkey, Georgia, Azerbaijan Seek to Develop Ties" (2012), *Anatolian Agency*, 8 June, http://w3.nexis.com/new/results/docview/docview.do?docLinkInd=true&risb=21_T15042759992&format=GNBFI&sort=BOOLEAN&startDocNo=1&resultsUrlKey=29_T15042763902&cisb=22_T15042763901&treeMax=true&treeWidth=0&csi=10962&docNo=13

"Turkey Completes a Major Step in the Marmaray Project" (2008), *Eurasia Daily Monitor*, 5:197, http://www.jamestown.org/single/?no_cache=1&tx_ttnews[tt_news]=34023, last accessed 9 April 2009.

"Turkey Not to Resume Ties with Armenia Until Karabakh 'Occupation' Ends" (2012), *Anatolian Agency*, 10 June, http://w3.nexis.com/new/results/docview/docview.do?docLinkInd=true&risb=21_T15043076872&format=GNBFI&sort=BOOLEAN&startDocNo=1&resultsUrlKey=29_T15043076876&cisb=22_T15043076875&treeMax=true&treeWidth=0&csi=10962&docNo=3.

"Turkey, Russia to Expand Blue Stream Pipeline" (2009), *Today's Zaman*, 28 March, http://www.todayszaman.com/tz-web/detaylar.do?load=detay&link=170826, last accessed 2 April 2009.

"Turkey Sends Two Military Observers to South Ossetia as Part of OSCE Mission" (2008), *Anatolian Agency*, 4 September, http://w3.nexis.com/new/results/docview/docview.do?docLinkInd=true&risb=21_T15048463066&format=GNBFI&sort=BOOLEAN&startDocNo=1&resultsUrlKey=29_T15048463070&cisb=22_T15048463069&treeMax=true&treeWidth=0&csi=10962&docNo=2.

"Turkey Suspends Nabucco Talks with Gaz de France" (2007), Reuters, 5 April, http://uk.reuters.com/article/oilRpt/idUKANK00020320070405, last accessed 31 March 2009.

"Turkey-UK Strategic Partnership" (2007), United Kingdom, Treaty, 25 October. http://www.fco.gov.uk/resources/en/news/2007/10/fco_not_251007_turkukstratpart, last accessed 19 March 2009.

"Turkey Would Review Nabucco if EU Talks Blocked" (2009), Reuters, January 19, http://in.reuters.com/article/oilRpt/idINLJ53530620090119, last accessed 31 March 2009.

"Turkey's Caucasian Adventure" (2000), *Mideast Mirror*, 2 February, http://w3.nexis.com/new/results/docview/docview.do?docLinkInd=true&risb=21_T15040008591&format=GNBFI&sort=BOOLEAN&startDocNo=101&resultsUrlKey=29_T15040014912&cisb=22_T15040014911&treeMax=true&treeWidth=0&csi=142710&docNo=111.

"Turkey's Caucasus Stability Initiative Driven by Own Interests" (2009), *Azerbaijan Press Agency*, 15 September, supplied by BBC Worldwide Monitoring, http://w3.nexis.com/new/results/docview/docview.do?docLinkInd=true&risb=21_T15006023227&format=GNBFI&sort=BOOLEAN&startDocNo=1&resultsUrlKey=29_T15006023231&cisb=22_T15006023230&treeMax=true&treeWidth=0&csi=10962&docNo=2

"Turkey's Energy Strategy" (2008), Directorate General for Energy, Water and Environment of Turkey, February.

"Turkey's Energy Strategy" (2010), *Turkish Government News*, 6 July, http://w3.nexis.com/new/results/docview/docview.do?docLinkInd=true&risb=21_T15040008591&format=GNBFI&sort=BOOLEAN&startDocNo=51&resultsUrlKey=29_T15040014912&cisb=22_T15040014911&treeMax=true&treeWidth=0&csi=372198&docNo=63.

"Turkish Defense Minister Hopes for Better Ties with Greece After Quake" (1999), *Anatolian Agency*, 28 August, http://w3.nexis.com/new/results/docview/docview.do?docLinkInd=true&risb=21_T15043007686&format=GNBFI&sort=BOOLEAN&startDocNo=1&resultsUrlKey=29_T15043007690&cisb=22_T15043007689&treeMax=true&treeWidth=0&csi=10962&docNo=15.

"Turkish Energy Corridor," Platts, http://www.platts.com/Natural%20Gas/Resources/News%20Features/turkey/europe.xml, last accessed 25 March 2009.

"Turkish, EU Leaders view Turkey's role in European Energy Policy" (2007), Anatolia Turkish News Agency, 5 June.

"Turkmenistan and Uzbekistan: Energy Profile" (2007), *Spero News*, 28 November, http://www.speroforum.com/site/article.asp?idCategory=35&idarticle=12534, last accessed 8 April 2009.

"Turkmenistan Energy Profile" (2009), Energy Information Administration homepage, http://tonto.eia.doe.gov/country/country_energy_data.cfm?fips=TX, last accessed 7 April 2009.

"Turkmenistan Needs $5 Billion Annually to Stem Output Drop,'" (2009), *Georgian Daily*, 18 March, http://georgiandaily.com/index.php?option=com_content&task=view&id=10583&Itemid=77, last accessed 16 April 2009.
"Turkmenistan's natural gas: mixed blessing" (2007), *Christian Science Monitor*, 15 May, http://www.csmonitor.com/2007/0515/p07s01-wosc.html?page=3, last accessed 9 April 2009.
"Ukraine-EU Summit Declarations" (2008), European Commission and Government of Ukraine, Kiev, Ukraine, 9 September.
"Ukrainian Foreign Minister Outlines Priorities in Relations with EU, NATO" (2008), BBC Monitoring International, 24 February.
"UN forum aims to spur investment in Asian Highway Network," UN homepage, http://www.un.org/apps/news/story.asp?NewsID=24690&Cr=&Cr1=, last accessed 17 April 2009.
"Updated 2- Turkey Vows Never to Use Gas Pipeline as a Weapon" (2009), Reuters, 19 January, http://uk.reuters.com/article/oilRpt/idUKLJ65769920090119?sp=true, last accessed 31 March 2009.
"US Ex-Im Approves Loan for Construction of Azerbaijani Satellite Azerspace" (2011), *Trend*, 29 April, http://en.trend.az/capital/it/1868667.html.
"Uzbekistan—The Oil and Gas Potential" (2002), *All Business*, 28 October, http://www.allbusiness.com/mining/oil-gas-extraction-crude-petroleum-natural/314931-1.html, last accessed on 18 March 2009.
"V ramkakh proekta TRACECA Kazakhstan planiruet realizovat' okolo 80 proektov na $30 milliardov" (Kazakhstan Plans to Implement about 80 Projects for $30 Billion within the TRACECA Project) (2007), *Panorama*, Kazakhstan, 14 December.
"Vilnius Conference Calls on EU to Pay More Attention to Energy Policy Coordination, Relations with Russia" (2007), Baltic News Service, 10 October.
"Vilnius Summit: Institutionalising a Process" (2007), Vilnius Conference Documents, (2007), BNS-Oct-9-11, Vilnius, 12 October.
"What Does France Understand from Privileged Partnership?" (2006), *Turkish Daily News*, January 17.
"White Stream Gas Pipeline Double Profitable for Azerbaijan: Ukraine's President" (2008) http://capital.trend.az/index.shtml?show=news&newsid=1211915&lang=EN, last accessed 1 April 2009.
"World Bank: Willing to Give Nabucco Financial Support" (2009), Thomson Reuters UK.
"Yulia Tymoshenko: Energy Cooperation is One of Key [*sic*] Priorities Between Ukraine and EU" (2008), web portal of Ukrainian government, 29 January, http://www.kmu.gov.ua/control/en/publish/article?art_id=108400364&cat_id=2297108, last accessed 1 April 2009.

Secondary Sources

Abbasov, Irshad, and Elnur Ahmadov (2010). "Mainstreaming Climate Change Impacts of Construction of 59-km long Masalli-Astara Highway Project," Bankwatch: NGO Forum on the Asian Development Bank.
Acharya, Amitav (2009). *Constructing a Security Community in Southeast Asia: ASEAN and the Problem of Regional Order* (New York: Routledge)
Adams, Terry (2009). "Baku Oil Diplomacy and 'Early Oil' 1994–1998: An External Perspective," in Fariz Ismailzade and Alexandros Petersen (eds.), *Azerbaijan in Global Politics: Crafting Foreign Policy* (ADA: Baku).
Adler, Emanuel, and Michael Barnett (1998). *Security Communities* (New York: Cambridge University Press).
Aktar, Cengiz (2006). "What does France understand from Privileged Partnership?" *Hurriyet, Turkish Daily News*, 17 January.
Aliriza, Bulent (2008). "Turkey and the Crisis in the Caucasus," *CSIS Commentary*, 9 September.

Allison, Roy (2004a). "Introduction," *International Affairs,* 80, 3, pp. 423–427.
Allison, Roy (2004b). "Regionalism, regional structures and security management in Central Asia," *International Affairs,* 80, 3, pp. 463–483.
Allison, Roy (2008). "Virtual regionalism, regional structures and regime security in Central Asia," *Central Asian Survey,* 27, 2.
Altstadt, Audrey L. (1997). "Azerbaijan's Struggle Toward Democracy," in Karen Dawisha and Bruce Parrott (eds.), *Conflict, Cleavage, and Change in Central Asia and the Caucasus* (New York: Cambridge University Press, 1997), pp. 143–145.
Amirahmadi, Hooshang (2000). "Challenges of the Caspian Region," in Hooshang Amirahmadi (ed.), *The Caspian Region at a Crossroad: Challenges of a New Frontier of Energy and Development* (New York: St. Martin's Press), pp. 16–17.
Andren, Nils Bertel Einer, and Karl E. Birnbaum (1976). *Beyond Detente: Prospects for East-West Cooperation and Security in Europe* (Leiden: Brill).
Aras, Bulent, and George Foster (1999). "Turkey: Looking for Light at the End of the Caspian Pipeline" in Michael Croissant and Bulent Aras (eds.), *Oil and Geopolitics in the Caspian Sea Region* (Westport: Praeger), pp. 234–235.
Asmus, Ronald, ed. (2006). *Next Steps in Forging a Euroatlantic Strategy for the Wider Black Sea* (Washington DC: German Marshall Fund).
Austrian Energy Agency on the Energy Community Treaty and related issues, http://www.eva.ac.at/enercee/enlargement.htm, last accessed 23 March 2009.
Aves, Jonathan (1996). "Post-Soviet Transcaucasia," in Roy Allison (ed.), *Challenges for the Former Soviet South* (Washington: Brookings Institution Press), pp. 163–165.
Axelrod, Robert, and Robert O. Keohane (1993). "Achieving Cooperation Under Anarchy: Strategies and Institutions," in *Neorealism and Neoliberalism: The Contemporary Debate* (New York: Columbia University Press), pp. 109–110.
Aydin, Mustafa (2005). "Regional Cooperation in the Black Sea and the Role of Institutions," *Perceptions,* 10, 3, pp. 57–84.
Babali, Tuncay (2005). "Implications of the Baku-Tbilisi-Ceyhan Main Oil Pipeline Project." *Perceptions* 10, 3, pp. 29–60.
Babali, Tuncay (2009). "Who's Losing Turkey?," *Huffington Post,* 17 November 2009.
Bagaturia, George, and Teimuraz Gorshkov (2001). "TRACECA – Restoration of Silk Route," *Japan Railway & Transport Review* 28, September.
Bakradze, David (2008), "'Georgia's European Way," remarks at "European Neighborhood Policy Action Plan Implementation," Tblisi, 5 February.
Bal, Idris (2000). *Turkey's Relations with the West and the Turkic Republics: The Rise and Fall of the "Turkic Model"* (Burlington, VT: Ashgate), pp. 205–206.
Baldwin, David A. (1993). "Neoliberalism, Neorealism, and World Politics," in David Baldwin (ed.), *Neorealism and Neoliberalism: The Contemporary Debate* (New York: Columbia University Press), p. 13.
Ball, George W. (1991). "Introduction," in Douglas Brinkley and Clifford Hackett (eds.), *Jean Monnet: The Path to European Unity* (New York: St. Martin's Press).
Baran, Zeyno (2007). "EU Energy Security: Time to End Russian Leverage." *Washington Quarterly,* 30, 4, pp. 131–144.
Barbieri, Katherine (2005). *The Liberal Illusion: Does Trade Promote Peace?* (Ann Arbor: University of Michigan Press).
Barysch, Katinka (2007), "Turkey's role in European energy security," London: Centre for European Reform. 12 December.
Baun, Michael (1995). "The Maastricht Treaty As High Politics: Germany, France, and European Integration," *Political Science Quarterly,* 110, 4.
Bellamy, Alex J. (2004). *Security Communities and Their Neighbors: Regional Fortresses or Global Integrators?* (New York: Palgrave Macmillan).
Bennett, Juliette (2002). "Multinational Corporations, Social Responsibility, and Conflict," *Journal of International Affairs,* 55, 2.
Berger, Lee (2001). "Proposed Legal Structure for the SilkSat Satellite Consortium: A Regional Intergovernmental Organization to Improve Telecommunications Infra-

structure in Central Asia and the Trans-Caucasus Region," *Law and Policy in International Business* 33, pp. 100ff.

Berman, Ilan (2004). "The New Battleground: Central Asia and the Caucasus," *Washington Quarterly*, 28, 1, pp. 59–69.

Bialer, Seweryn (2001). *Stalin's Successors: Leadership, Stability, and Change in the Soviet Union* (New York: Cambridge University Press).

Bicchi, Frederica (2006). "'Our size fits all': normative power Europe and the Mediterranean," *Journal of European Public Policy*, 13, 2, pp. 286–303.

The Black Sea Between the EU and Russia: Security, Energy, Democracy (2006). (Bergedorfer Gesprachskreis).

Blythe, Stephen E. (2007). "Azerbaijan's E-Commerce Statutes: Contributing to Economic Growth and Globalization in the Caucasus Region," *Columbia Journal of East European Law* 1, p. 44.

Borzel, Tanja, and Yasemin Pamuk (2012). "Pathologies of Europeanisation: Fighting Corruption in the Southern Caucasus," *West European Politics*, 35, 1, pp. 79–97.

BP (2005). "BTC/SCP Commemorate a Programme to Help Combat HIV/AIDS in Azerbaijan," 5 December 2005, http://www.bp.com/genericarticle.do?categoryId=9006615&contentId=7053922.

BP (2007). "BTC and SCP Open Two Water Purification Facilities in Kurdamir," 1 February 2007, http://www.bp.com/genericarticle.do?categoryId=9006615&contentId=7028200.

BP (2008). "Azerbaijan Sustainability Report 2008," http://www.bp.com/liveassets/bp_internet/bp_caspian/bp_caspian_en/STAGING/local_assets/downloads_pdfs/s/BP_Azerbaijan_Sustainability_Report_2008_eng_final.pdf.

BP (2008). *BP Statistical Review of World Energy June 2008*.

BP Georgia. "BP Pledge," http://www.bpgeorgia.ge/go/doc/1339/145018/.

BP Press Release (2007). "Azerbaijan Government and BP Sign Bilateral Security Protocol," November 13, 2007, http://www.bp.com/genericarticle.do?categoryId=9006615&contentId=7038375.

BP. "BP Azerbaijan Sustainability Report: 2008," http://www.bp.com/liveassets/bp_internet/bp_caspian/bp_caspian_en/STAGING/local_assets/downloads_pdfs/s/BP_Azerbaijan_Sustainability_Report_2008_eng_final.pdf.

BP. "Engaging with Communities along Pipeline Routes," http://www.bp.com/sectiongenericarticle.do?categoryId=9028362&contentId=7051665.

BP. "External Monitoring of Our Activities," http://www.bp.com/subsection.do?categoryId=9006625&contentId=7013426.

BP. "Security and Continuity," http://www.bp.com/sectiongenericarticle.do?categoryId=9028235&contentId=7051416.

BP. "Security and Human Rights," http://www.bp.com/sectiongenericarticle.do?categoryId=9027869&contentId=7050848.

Brown, Gordon, and Recep Tayyip Erdogan, (2007). "Turkey-UK Strategic Partnership," August.

Brown, Rupert (2000). *Group Processes* (Malden, MA: Blackwell Publishers).

Bryza, Matthew J. (2006). "The Policy of the United States toward the Black Sea Region," in Ronald Asmus (ed.), *Next Steps in Forging a Euroatlantic Strategy for the Wider Black Sea* (Washington DC: German Marshall Fund).

BSEC (1999). *Charter of the Organization of the Black Sea Economic Cooperation* (Istanbul: BSEC).

Burgess, Michael (1996). "Introduction: Federalism and Building the European Union," *Publius* 26, 4.

Busse, Niklas (1999). "Constructivism and Southeast Asian Security," *The Pacific Review* 12, 1, pp. 39–60.

Butler, Darren, and Paul Taylor (2007). "EU, Turkey at Odds over Energy Community," Reuters Istanbul, 7 June, http://uk.reuters.com/article/oilRpt/idUKL0575347620070605, last accessed 25 March 2009

Buzan, Barry (1991). *People, States and Fear: An Agenda for International Security Studies in the Post–Cold War Era* (Boulder: L. Rienner).

Cagaptay, Soner (2009). "Is Turkey Leaving the West?," *Foreign Affairs*, 26 October.

Cantori, Louis J., and Stephen L. Spiegel (1973). "The Analysis of Regional International Politics: The Integration Versus the Empirical Systems Approach," *International Organization* 27, 4, p. 476.

Caspian Development Advisory Panel (2007). "Final report and Conclusions," Jan. 2007, http://www.bp.com/liveassets/bp_internet/bp_caspian/bp_caspian_en/STAGING/local_assets/downloads_pdfs/c/CDAP_Final_Report_and_Conclusions_en.pdf.

Caspian Revenue Watch (2003). *Caspian Oil Windfalls: Who Will Benefit?* (New York: Open Society Institute).

Celac, Sergiu, Michael Emerson, and Nathalie Tocci (2000). "Annex G: Remarks by Eduard Shevardnadze, President of Georgia at the Informal Summit of the Black Sea Economic Cooperation Istanbul, November 17, 1999," in *A Stability Pact for the Caucasus: A Consultative Document of the CEPS Task Force on the Caucasus* (Brussels: Center for European Policy Studies), pp. 73–74, http://aei.pitt.edu/11697/1/55.pdf.

Chung, Chien-peng (2004). "The Shanghai Co-operation Organization: China's Changing Influence in Central Asia," *China Quarterly*, 180, pp. 1000–1001.

Clapham, Christopher (2000). *Africa and the International System: The Politics of State Survival* (New York: Cambridge University Press).

Clemens, Walter C. (1990). *Can Russia Change? The USSR Confronts Global Interdependence* (Boston: Unwin Hyman).

Coffey International Development (2009). *Validation of the Extractive Industries Transparency Initiative (EITI) in the Republic of Azerbaijan.*

Cohen, Ariel (2002). "Iran's Claim Over Caspian Sea Resouces Threaten Energy Security," *Heritage Backgrounder*, 5 September.

Commission of the European Communities (2007a), *A Stronger European Neighbourhood Policy* (Brussels: Commission of the European Communities).

Commission of the European Communities (2007b), *Black Sea Synergy: A New Regional Cooperative Initiative* (Brussels: Commission of the European Communities).

Cooley, Alexander (2008). "Principles in the Pipeline: Managing Transatlantic Values and Interests in Central Asia," *International Affairs* 84, 6, (2008), p. 1184.

Copeland, Dale C. (2006). "The Constructivist Challenge to Structural Realism: A Review Essay," in Stefano Guzzini and Anna Leander (eds.), *Constructivism and International Relations: Alexander Wendt and His Critics* (New York: Routledge).

Cornell, Svante (1998). "Turkey and the Conflict in Nagorno Karabakh: A Delicate Balance," *Middle Eastern Studies* 34, 1, p. 51.

Creamer, Martin (2010). "Social Risk Mitigation Taking Mining's Centre Stage," *Mining Weekly*, 22 January 2010.

Crowley, Patrick M. (2004). *Crossing the Atlantic: Comparing the European Union and Canada* (Surrey, UK: Ashgate).

Daly, John (2007), "New Gas Pipeline to Link Azerbaijan and Italy, Via Turkey" *Eurasia Daily Monitor* (Jamestown Foundation).

Dannreuther, Roland (2006). "Developing the Alternative to Enlargement: The European Neighbourhood Policy," *European Foreign Affairs Review*, 11, p. 184.

De Waal, Thomas (2005). "Georgia and its Distant Neighbors," in Bruno Coppieters and Robert Legvold (eds.), *Statehood and Security: Georgia after the Rose Revolution* (Cambridge, MA: MIT Press), pp. 314–315, 322–325.

De Waal, Thomas (2010). *The Caucasus: An Introduction* (Oxford: Oxford University Press), pp. 7–8.

"Debordering and Rebordering in the Mediterranean," *EUI Working Papers* 43 (Florence: European University Institute).

Demchenko, I. O., and P. D. Kaminsky (2008), "TRASEKA kak tsel' ekonomicheskogo sotrudnichestva" (TRACECA as a Goal of Economic Collaboration), Donetsk National University of Economics and Trade, http://www.rusnauka.com/28_NIOXXI_2008/Economics/35420.doc.htm, accessed on 16 March 2009.

Demirag, Yelda (2004). "EU Policy towards South Caucasus and Turkey," *Perceptions: Journal of International Affairs* (Winter), p. 92.

Denisenko, A. A. (2007). "Problemy ispol'zovaniia mezhdunarodnogo koridora 'TRASEKA'" (Problems of Using the International TRACECA Corridor), Presentation of Deputy General Secretary of KazATO A.A. Denisenko at a roundtable dedicated to "State and Private Sector Partnership to Solving Transportation Problem of Countries with No Access to Sea," Batumi, Georgia, 19 May 2007, http://www.iru-cis.ru/iru-moscow/2007/05_batumi/doc/Denisenko-Batumi_TransCorridors_-Rus.pdf, accessed 16 March 2009.

Deutsch, Karl W. (1968). *The Analysis of International Relations* (Englewood Cliffs, NJ: Prentice-Hall).

Deutsch, Karl W. et al. (1957). *Political Community in the North Atlantic Area: International Organization in the Light of Historical Experience* (Princeton: Princeton University Press).

Deverakonda, Puja (2008). "Gaz de France Cancels Nabucco Interest," http://www.ypfp.org/content/gaz-de-france-cancels-nabucco-interest, last accessed 31 March 2009.

Dwan, Renata (2000). "Introduction" in Renata Dwan and Oleksandr Pavliuk (eds.), *Building Security in the New States of Eurasia: Subregional Cooperation in the Former Soviet Space* (New York: M. E. Sharpe).

Elder, Miriam (2007). *St. Petersburg Times*, 15 May.

Energy Sector Management Assistance Programme (ESMAP) (2003). *Cross-Border Oil and Gas Pipelines: Problems and Prospects*. Washington DC: Joint World Bank Group and United Nations Development Programme.

European Union Commission on Enlargement. "EU-Turkey Relations." European Union. http://ec.europa.eu/enlargement/candidate-countries/turkey/eu_turkey_relations_en.htm, last accessed 19 March 2009.

Fawcett, Louise (2004). "Exploring regional domains: a comparative history of regionalism," *International Affairs*, 80, 3, pp. 429–446.

Feldstein, Helen S. (1967). "A Study of Transaction and Political Integration: Transnational Labour Flow within the European Economic Community," *Journal of Common Market Studies* 6, 1, p. 24-55.

Ferrero-Waldner, Benita (2006). "Opening Speech" (speech at the conference "Towards an EU External Energy Policy to Assure a High Level of Supply Security," Brussels, 20 November.

Ferrero-Waldner, Benita (2007). "European External Relations and Energy Policy: Towards an International Energy Strategy" (speech, Energy Conference, Cologne, 11 September).

Finger, J. Michael (2004). "Introduction," in J. Michael Finger and Philip Schuler (eds.), *Poor People's Knowledge: Promoting Intellectual Property in Developing Countries* (Washington DC: World Bank).

Franke, Benedikte (2008). "Africa's Evolving Security Architecture and the Concept of Multilayered Security Communities," *Cooperation and Conflict: Journal of the Nordic International Studies Association* 43, 3, pp. 313–340.

Friis, Lykke, and Anna Murphy (1999). "The European Union and Central and Eastern Europe: Governance and Boundaries," *Journal of Common Market Studies* 37, 2, pp. 211–232.

Gachechiladze, Revaz (2002). "Geopolitics in the South Caucasus: Local and External Players," *Geopolitics* 7, 1, pp. 113–118, 127.

Gebhard, Carmen (2007). *Assessing EU Actorness Towards its 'Near Abroad': The European Neighbourhood Policy* (Maastricht: European Institute of Public Administration).

Gegeshidze, Archil (2002). "Georgia's Function as a Transit Country and Sustainable Development," World Summit on Sustainable Development.

Geropoulos, Kostis (2008). "Tymoshenko Puts New White Stream Pipeline on EU Table," *European Weekly*, http://www.neurope.eu/articles/82525.php, last accessed 1 April 2009.
Ghafouri, Mahmoud (2008). "The Caspian Sea: Rivalry and Cooperation." *Middle East Policy* 15, 2, pp. 81–96.
Gillingham, John (1991). "Jean Monnet and the European Coal and Steel Community: A Preliminary Appraisal," in Douglas Brinkley and Clifford Hackett (eds.), *Jean Monnet: The Path to European Unity* (New York: St. Martin's Press).
Godzimirski, Jakub M. (2009). "Russia's Energy Strategy and Prospects for a Northern Dimension Energy Partnership," in Pami Aalto, Helge Blakkisrud, and Hanna Smith (eds.), *The New Northern Dimension of the European Neighbourhood* (Brussels: Center for European Policy Studies).
Goodby, James E., Petrus Buwalda, and Dmitri Trenin (2002). *A Strategy for Stable Peace: Toward a Euroatlantic Security Community* (Washington DC: U.S. Institute for Peace Press).
Gorshkov, Teimuraz, and George Bagaturia (2000). "Developments at Georgian Railway," *Japan Railway and Transport Review* 24, pp. 42–43.
Greico, Joseph M. (1988). "Anarchy and the Limits of Cooperation: A Realist Critique of the Newest Liberal Institutionalism," *International Organization* 42, 3, pp. 485–507.
Grigoriadis, Ioannis N., (2008). "Natural Gas Corridors in Southeastern Europe and European Energy Security," ELIAMEP Thesis, July.
GUAM (2001). *Charter of Organization for democracy and economic development—GUAM* (Kyiv: GUAM).
Guler, Hilmi (2007). Keynote Speech to the 3rd Bosphorus Conference, Istanbul, October 6.
Gulmammadov, Rufat H. (2009). "The National ICT Strategy and Key Indicators on Information Society of Azerbaijan Republic," e-Govshare Conference.
Haas, Ernst B. (1958). "The Challenge of Regionalism," *International Organization* 12, 4, pp. 440–458.
Haas, Ernst B. (1970). "The Study of Regional Integration: Reflections on the Joy and Anguish of Pretheorizing," *International Organization* 24, 4, pp. 607–646
Haas, Ernst B. (1961). "International Integration: The European and Universal Process," *International Organization*, 15.
Haas, Ernst B. (1968). *The Uniting of Europe: Political, Social, and Economic Forces, 1950–1957* (Notre Dame: University of Notre Dame Press).
Haas, Ernst B. (1971). "The Study of Regional Integration: Reflections on the Joy and Anguish of Pretheorizing," in Lindberg and Scheingold (eds.), *Regional Integration: Theory and Research* (Cambridge, MA: Harvard University Press), pp. 618–619.
Haas, Ernst B. (1975). *The Obsolescence of Regional Integration Theory* (Berkeley: University of California Institute of International Studies).
Haas, Ernst B., and Philippe Schmitter (1964). "Economics and Differential Patterns of Political Integration: Projections about Unity in Latin America," *International Organization* 18,4, p. 710.
Hackett, James, ed. (2008). *The Military Balance* (London: Routledge).
Hadjian, Avedis Bedros (2001). "Azerbaijan's Energy Policy and Its Implications for Russian Security," in Bulent Gokay (ed.), *The Politics of Caspian Oil* (New York: Palgrave), pp. 131–133, 135–136, 139–140.
Hakobyan, Grigor (2005). "Armenia Evaluates the Pros and Cons of the BTC Pipeline," *CACI Analyst*, 15 June.
Hancock, Kathleen (2006). "Escaping Russia, Looking to China: Turkmenistan Pins Hopes on China's Thirst for Natural Gas," *China and Eurasia Forum Quarterly* 4, 3, pp. 78–79.
Harmes, Adam (2005). *The Return of the State: Protesters, Power-Brokers, and the New Global Compromise* (Vancouver: Douglas and McIntyre).

Helly, Damien, and Giorgi Gogia, "Georgian Security and the Role of the West," in Bruno Coppieters and Robert Legvold (eds.), *Statehood and Security: Georgia after the Rose Revolution* (Cambridge, MA: MIT Press), pp. 270–281.
Hemmer, Christopher, and Peter J. Katzenstein (2002). "Why is there no NATO in Asia? Collective Identity, Regionalism, and the Origins of Multilateralism," *International Organization* 56, 3, p. 576.
Hettne, Bjorn, Andras Inotai, and Osvaldo Sunkel (eds.) (2001). *Comparing Regionalisms: Implications for Global Development* (New York: Palgrave).
Hill, Fiona (2003). "Seismic Shifts in Eurasia: The Changing Relationship between Turkey and Russia and its Implications for the South Caucasus," *Journal of Southeast European and Black Sea Studies*, 3, 3, pp. 66, 70.
Hill, Fiona, and Omer Taspinar (2006). "Turkey and Russia: Axis of the Excluded?" *Survival* 48, 1, pp. 84, 87.
Hill, Jeffrey (2012). "Azercosmos: Work Underway on New Azerspace Satellite," *Satellite Today*, 10 April, http://www.satellitetoday.com/twitter/Azercosmos-Work-Underway-on-New-Azerspace-Satellite_38623.html.
Hishmeh, George S. (1999). "Letter from America," *Mideast Mirror*, 16 April, http://w3.nexis.com/new/results/docview/docview.do?docLinkInd=true&risb=21_T15040008591&format=GNBFI&sort=BOOLEAN&startDocNo=101&resultsUrlKey=29_T15040014912&cisb=22_T15040014911&treeMax=true&treeWidth=0&csi=142710&docNo=114.
Hofer, Stephen D. (2007). *Neo-functionalism reloaded. The Energy Community of Southeast Europe*, prepared for the 9th Annual Kokkalis Graduate Student Workshop, Harvard University (Vienna: Universität Wien).
Hoffman, David I. (2000). "Azerbaijan: The Politicization of Oil," in Robert Ebel and Rjan Menon (eds.), *Energy and Conflict in Central Asia and the Caucasus* (New York: Rowman & Littlefield), pp. 58–59, 60–63, 65–67.
Hoffman, Stanley (1964). "The European Process at Atlantic Crosspurposes," *Journal of Common Market Studies*, 2, 3, pp. 85–101.
Hoffmann, Stanley (1966). "Obstinate or Obsolete? The Fate of the Nation-State and the Case of Western Europe," *Daedalus*, 95, 3.
Hoffmann, Stanley (1982). "Reflections on the Nation State in Western Europe Today," *Journal of Common Market Studies*, vol. 21.
Holsti, Kalevi (1996). *The State, War and the State of War* (Cambridge Studies in International Relations).
Hough, Peter (2008). *Understanding Global Security* (New York: Routledge).
Howard, Glen E. (1998). "NATO and the Caucasus: The Caspian Axis," in Stephen J. Blanks (ed.), *NATO After Enlargement: New Challenges, New Missions, New Forces* (Carlisle Barracks, Pennsylvania), p. 152.
Human Rights Watch (2004). *Crushing Dissent: Repression, Violence, and Azerbaijan's Elections* (London: HRW).
Human Rights Watch (2007). *Crossing the Line: Georgia's Violent Dispersal of Protestors and Raid on Imedi Television* (London: HRW).
Human Rights Watch (2008). *Azerbaijan's Human Rights Situation Prior to the Elections* (London: HRW).
Human Rights Watch (2009). *Turkey: Events of 2008* (London: HRW)
Huntington, Samuel P. (1996). *The Clash of Civilizations and the Remaking of World Order* (Simon and Schuster: London).
Ignatieff, Michael (1994). *Blood and Belonging* (London: Vintage).
Independent Evaluation Group (2007). *Georgia Country Assistance Evaluation Approach Paper*. (Washington DC: World Bank Group).
Independent Evaluation Group (2008). *Doing Business: An Independent Evaluation Taking the Measure of the World Bank-IFC Doing Business Indicators* (Washington DC: World Bank Group).
"Integrating EU and Turkish Foreign Policy," Brussels: CEPS, http://shop.ceps.be/downfree.php?item_id=1143, last accessed 25 March 2009.

International Bank for Reconstruction and Development (IBRD) (2009). *Global Economic Prospects 2009: Commodities at the Crossroads* (Washington DC: World Bank Group).
International Bank for Reconstruction and Development (IBRD) (2008). *Doing Business 2009: Europe and Central Asia* (Washington DC: World Bank Group).
International Bank for Reconstruction and Development (IBRD) and International Development Association (IDA) (2008). *Status of Projects in Execution Report for FY2008* (Washington DC: World Bank Group).
International Bank for Reconstruction and Development (IBRD) and International Development Association (IDA) (2008). *Status of Projects in Execution for FY 2008 Europe and Central Asia Region Country: Georgia* (Washington DC: World Bank Group).
International Bank for Reconstruction and Development (IBRD) and International Finance Corporation (IFC) (2005). *Country Assistance Strategy Progress Report for the Republic of Turkey for the Period FY 2004–2007.* (Washington DC: World Bank Group).
International Bank for Reconstruction and Development (IBRD), International Development Association (IDA), and International Finance Corporation (IFC) (2006). *Country Partnership Strategy for Republic of Azerbaijan* (Washington DC: World Bank Group).
International Bank for Reconstruction and Development (IBRD), International Development Association (IDA), and International Finance Corporation (IFC) (2008). *Country Partnership Strategy Progress Report for Azerbaijan for the Period FY 2007–2010* (Washington DC: World Bank Group).
International Bank for Reconstruction and Development (IBRD), International Finance Corporation (IFC), and Multilateral Investment Guarantee Agency (MIGA) (2008). *Country Partnership Strategy with the Republic of Turkey for the Period FY 2008–2011* (Washington DC: World Bank Group).
International Development Association (IDA) and International Finance Corporation (IFC) (2005). *Country Partnership Strategy for Georgia* (Washington DC: World Bank Group).
International Development Association (IDA) and International Finance Corporation (IFC) (2008). *Country Partnership Strategy Progress Report for Georgia for the Period FY 2006–2009* (Washington DC: World Bank Group).
International Finance Corporation (IFC) (2004). *Sustainability Report 2004* (Washington DC: World Bank Group).
International Finance Corporation (IFC) (2006). *Doing Business 2007: How to Reform* (Washington DC: World Bank Group).
International Monetary Fund (IMF) (2004). *Azerbaijan Republic: Poverty Reduction Strategy Paper Progress Report* (Washington DC: International Monetary Fund).
International Monetary Fund (IMF) (2007). *Republic of Azerbaijan: Staff Report for the 2007 Article IV Consultation* (Washington DC: International Monetary Fund).
International Monetary Fund (IMF) (2007). *Turkey: Staff Report for the 2007 Article IV Consultation.* (Washington DC: International Monetary Fund).
International Monetary Fund (IMF) (2008). *Republic of Azerbaijan: Staff Report for the 2008 Article IV Consultation* (Washington DC: International Monetary Fund).
International Monetary Fund (IMF) (2009). *Georgia: Staff Report for the 2009 Article IV Consultation and Second Review Under the Stand-By Arrangement* (Washington DC: International Monetary Fund).
Ismailzade, Fariz (2007). "Regional Leaders Seal Kars-Akhalkalaki-Tbilisi Railway Deal," *Eurasia Daily Monitor* (Jamestown Foundation).
Ismayilov, Rovshan (2007). "Azerbaijan, Georgia, and Turkey: Building a Transportation Triumvirate?" *EurasiaNet*, 2 June.
Jawad, Pamela (2008). "Conflict Resolution Though Democracy Promotion? The Role of the OSCE in Georgia," *Democratization*, 15, 3, p. 618.
Jervis, Robert (1999). "Realism, Neoliberalism, and Cooperation: Understanding the Debate," *International Security*, 24, 1, pp. 42–63.

Jordan, Robert and Werner Feld (2001). *International Organizations: A Comparative Approach to the Management of Cooperation* (Westport, CT: Greenwood Publishing).

Jun, Zhan (1996). "Low Politics Versus High Politics: The Case of Beijing-Taipei Detente Since the Late 1980s," in *Dynamics and Dilemma: Mainland, Taiwan and Hong Kong in a Changing World* (Commack, NY: Nova Science).

Kahraman, Sevilay (2005). "The European Neighbourhood Policy: The European Union's New Engagement Towards Wider Europe," *Perceptions*, 10, 3, pp. 1–28.

Kakachia, Kornely K (2011). "Challenges to the South Caucasus Regional Security Aftermath of Russian-Georgian Conflict: Hegemonic Stability or New Partnership?" *Journal of Eurasian Studies*, 2, pp. 16–17.

Kalyuzhnova, Yelena (2006). "Overcoming the Curse of Hydrocarbon: Goals and Governance in the Oil Funds of Kazakhstan and Azerbaijan," *Comparative Economic Studies*, 48.

Kanbolat, Hasan (2007). "Caucasia Connects to Europe through Turkey: Kars-Tbilisi-Baku Railway Project," http://www.todayszaman.com/tz-web/yazarDetay.do?haberno=123002, last accessed 3 April 2009.

Kanbolat, Hasan (2008). "What is Caucasian stability and cooperation? What can Turkey do in the Caucasus?," *Today's Zaman*, http://www.todayszaman.com/newsDetail_getNewsById.action?load=detay&link=150578.

Karagiannis, Emmanuel (2002). *Energy and Security in the Caucasus* (New York: Routledge), pp. 144–146.

Kasim, Kamer (2001). "The Nagorno-Karabakh Conflict, Caspian Oil and Regional Powers," in Bulent Gokay (ed.), *The Politics of Caspian Oil* (New York: Palgrave), pp. 187–188, 190–191.

Katinka, Barysch (2007). "Turkey's Role in European Energy Security," *Centre for European Reform Policy Brief*, December.

Katzenstein, Peter J., Robert O. Keohane, and Stephen D. Krasner (1998). "International Organization and the Study of World Politics," *International Organization* 52, 4, p. 660.

Kaufman, Daniel, Aart Kraay, and Massimo Mastruzzi (2006). *Governance Matters V: Aggregate and Individual Governance Indicators for 1996–2005* (Washington DC: World Bank Group).

Kegley, Charles, and Eugene Wittkopf (1981). *World Politics: Trends and Transformation* (New York: St. Martin's Press).

Keohane, Robert O. (1986). "Theory of World Politics: Structural Realism and Beyond," in Robert O. Keohane (ed.), *Neorealism and Its Critics* (New York: Columbia University Press).

Keohane, Robert O. (1989). "Neoliberal Institutionalism: A Perspective on World Politics," in *International Institutions and State Power: Essays in International Relations Theory* (Boulder: Westview Press), p. 3.

Keohane, Robert O. (1989). *International Institutions and State Power: Essays in International Relations Theory* (Boulder: Westview Press).

Keohane, Robert O. (2005). *After Hegemony: Cooperation and Discord in the World Political Economy* (Princeton: Princeton University Press).

Keohane, Robert O., and Joseph S. Nye (1977). *Power and Interdependence: World Politics in Transition* (Boston: Little, Brown), p. 9.

Keohane, Robert O., and Lisa L. Martin (1995). "The Promise of Institutionalist Theory," *International Security* 20, 1, p. 40.

Kim, Joongi (2003). "Sub-regionalism, regionalism, trans regionalism. Implications for economic integration and international trade policies," *Asia Europe Journal*, 1, pp. 183–196.

King, Charles (2004). *The Black Sea: A History* (Oxford: Oxford University Press).

Kolstø, Pål, and Helge Blakkisrud (2012). "De facto states and democracy: The case of Nagorno-Karabakh," *Communist and Post-Communist Studies*, 45, 1–2 (Amsterdam: Elsevier, Ltd.), p. 1, http://ac.els-cdn.com/S0967067X12000141/1-s2.0-

S0967067X12000141-main.pdf?_tid=6c8d09457f9021e3dcb6b225a27ae720&acdnat=1341848267_cc88f2ece6710fa701c7c01054303d7c.

Konyrova, Kulpash (2007). "Putin Deal Torpedoes Trans-Caspian Gas Pipeline Plans," *New Europe*, 17 May, http://www.neurope.eu/view_news.php?id=73862, last accessed 1 April 2009.

Koolaee, Elaheh, and Mohammed Hossein Hafezian (2010). "The Islamic Republic of Iran and the South Caucasus Republics," *Iranian Studies* 43, 3, pp. 398–401.

Kramer, Andrew E. (2009). "New Gas Pipeline from Central Asia Feeds China," *New York Times*, 15 December.

Kucera, Joshua (2009). "The Nabucco Carrot and U.S.-Iran Engagment," *World Politics Review*, 5 May.

Kuhonta, Erik Martinez (2006). "Walking a Tightrope: Democracy Versus Sovereignty in ASEAN's Liberal Peace," *The Pacific Review*, 19, 3, pp. 337–358.

Kulieva, G. (2003). "Yubilei TRASEKA. Pervyi itogy novogo 'Shelkogo puti'" (TRACECA's Jubilee. First Results of the New "Silk Road"), *Ekho*, 29 June.

Laursen, Finn (2002). *Theories of European Integration*, background paper for lecture on 'European Integration: What and Why?' at Graduate Institute of European Studies, Tamkang University Taipei, Taiwan, March.

Lavenex, Sandra (2004). "EU external governance in 'wider Europe,'" *Journal of European Public Policy*, 11, 4, pp. 680–700.

Liao, Janet Xuanli (2006). "A Silk Road for Oil: Sino-Kazakh Energy Diplomacy," *Brown Journal of World Affairs*, 7, 2, p. 39.

Libaridian, Gerald J. (2004). "Armenia's Strategic Significance," *Modern Armenia: People, Nation, State* (New Brunswick, NJ: Transaction Publishers), pp. 96–97.

Lindberg, Leon (1963). *The Political Dynamics of European Economic Integration* (Stanford University Press).

Lindberg, Leon (1971). "Political Integration as a Multidimensional Phenomenon Requiring Multivariate Measurement," in Lindberg and Scheingold (eds.), *Regional Integration: Theory and Research* (Cambridge, MA: Harvard University Press), pp. 649–731.

Long, David, and Lucian Ashworth (1999). "Working for Peace: The Functional Approach, Functionalism and Beyond," in Ashworth and Long (eds.), *New Perspectives on International Functionalism* (New York: St. Martin's Press).

Luecke, Matthias, and Natalia Trofimenko (2008). "Whither Oil Money? Redistribution of Oil Revenue in Azerbaijan," in Boris Najman, Richard Pomfret, and Gael Raballand (eds.), *The Economics and Politics of Oil in the Caspian Basin: The Redistribution of Oil Revenues in Azerbaijan and Central Asia* (New York: Routledge), p. 137.

Lussac, Samuel (2008). "The Baku-Tbilisi-Kars Railroad and its Geopolitical Implications for the South Caucasus," *Caucasian Review of International Affairs*, 2, 4, pp. 212–224.

Lynch, Allen (2005). *How Russia Is Not Ruled: Reflections on Russian Political Development* (New York: Cambridge University Press).

MacFarlane, S. Neil (2004). "The United States and Regionalism in Central Asia," *International Affairs*, 80, 3, pp. 447–461.

MacMillan, Catherine (2009). "The Application of Neofunctionalism to the Enlargement Process: The Case of Turkey," *Journal of Common Market Studies*, 47, 4, p. 805.

Magen, Amichai (2006). "The Shadow of Enlargement: Can the European Neighbourhood Policy Achieve Compliance?," *Columbia Journal of European Law*, 12, 2, pp. 383–427.

Mankoff, Jeffrey (2009). "Eurasian Energy Security," *Council Special Report No. 43* (New York: Council on Foreign Relations).

Mansfield, Edward, and Brian Pollins (2003). "Interdependence and Conflict: An Introduction," in Mansfield and Pollins (eds.), *Economic Interdependence and International Conflict: New Perspectives on an Enduring Debate* (Ann Arbor: University of Michigan Press).

Mansfield, Edward D., and Jon C. Pevehouse (2003). "Institutions, Interdependence, and International Conflict," in Gerald Schneider, Katherine Barbieri, and Nils Petter Gleelitsch (eds.), *Globalization and Armed Conflict* (New York: Rowman & Littlefield).

Marketos, Thrassy N. (2009). *China's Energy Geopolitics: The Shanghai Cooperation Organization and Central Asia* (New York: Routledge), pp. 15–16.

Martin, Lisa L. (1992). "Interests, Power, and Multilateralism," *International Organization*, 46, 4, pp. 765–792.

Mayer, Sebastian (2006). "TACIS and EU's Security of Energy Supply: The Commission as a Strategic Actor in External Relations," *TransState Working Papers*, 46 (University of Bremen).

Mearsheimer, John J. (1995). "The False Promise of International Institutions," *International Security* 19, 3, p. 7.

Mehdiyeva, Nazrin (2011). *Power Games in the Caucasus: Azerbaijan's Foreign and Energy Policy Towards the West, Russia and the Middle East* (New York: I.B. Tauris), pp. 185–186.

Mitchell, Lincoln A. (2006). "Democracy in Georgia Since the Rose Revolution," *Orbis*, Fall, pp. 670, 672.

Mitra, Pradeep (2008). *Innovation, Inclusion and Integration: From Transition to Convergence in Eastern Europe and the Former Soviet Union* (Washington DC: World Bank Group).

Mitrany, David (1968). *A Working Peace System* (Chicago: Quadrangle Books).

Mojtahed-Zadeh, Pirouz (2000). "The Geo-Politics of the Caspian Region," in Hooshang Amirahmadi (ed.), *The Caspian Region at a Crossroad: Challenges of a New Frontier of Energy and Development* (St. Martin's Press: NY), pp. 180–181.

Moller, Frank (2007). *Thinking Peaceful Change: Baltic Security Policies and Security Community Building* (Syracuse: Syracuse University Press).

Moravcsik, Andrew (1999). "Is Something Rotten in the State of Denmark?: Constructivism and European Integration," *Journal of European Public Policy*, 6, 4, pp. 669–681.

Moravcsik, Andrew, and Frank Schimmelfennig (2009). "Liberal Intergovernmentalism," in Antje Wiener and Thomas Diez (eds.), *European Integration Theory* (New York: Oxford University Press), p. 72.

Morgan, Roger (1973). *High Politics, Low Politics: Toward a Foreign Policy for Western Europe* (Sage: Beverly Hills).

Morgenthau, Hans (1968). "Introduction," in *A Working Peace System* by David Mitrany (Chicago: Quadrangle Books).

Morris, Kathleen (2008). "The Emergence of Customary International Law Recognizing Corporate Liability for Violations of Human Rights and Environmental Law," *Gonzaga Journal of International Law*, 11.

Morse, Edward L. (1976). *Modernization and the Transformation of International Relations* (New York: The Free Press).

Moustakis, Fotios (2003). *The Greek-Turkish Relationship and NATO* (Portland, OR: Frank Cass), pp. 173–174.

Moustakis, Fotios, and Tracey German (2009). *Securing Europe: Western Interventions Towards a New Security Community* (New York: Tauris Academic).

Mutlu, Can E. (2011). "A De Facto Cooperation? The Increasing Role of the European Union in Improved Relations Between Georgia and Turkey," *Comparative European Politics*, 9, 4/5, p. 556.

Nathan, Laurie (2010). "The African Union and Regional Organisations in Africa: Communities of Insecurity," *African Security Review*, 19, 2, p. 106.

Nathan, Laurie (2006). "Domestic Instability and Security Communities," *European Journal of International Relations*, 12, 2, pp. 275–299.

National Energy Grid Map, Central Asia, Global Energy Network Institute homepage, http://www.geni.org/globalenergy/library/national_energy_grid/central-asia/graphics/cent_asia_elec_lg.gif, last accessed 9 April 2009.

Nezhinskaya, Anna (2007). Interview with General Secretary of TRACECA's Permanent Secretariat of the Inter-Governmental Commission Rustan Jenalinov,

"TPACEKA otkriyt dlya vsekh?" (TRACECA is Open to Everyone?), *RZD Partner*, 9 November 2007, http://www.rzd-partner.ru/interview/2007/09/11/310945.html, last accessed 16 March 2009.
Nichol, Jim (2007). "Armenia, Azerbaijan, and Georgia: Political Developments and Implications for U.S. Interests," *CRS Report for Congress* (Washington DC: Congressional Research Service).
Nichol, Jim (2008). "Georgia: Recent Developments and U.S. Interests," *CRS Report for Congress* (Washington DC: Congressional Research Service).
Nichol, Jim, Steven Woehrel, and Bernard Gelb (2007). "Russia's Cutoff of Natural Gas to Ukraine: Context and Implications," in Ulrich Nichol (ed.), *Focus on Politics and Economics of Russia and Eastern Europe* (Hauppage: Nova Publishing).
Niemann, Arne (1998). "The PHARE Programme and the Concept of Spillover: Neo-functionalism in the Making," *Journal of European Public Policy*, 5, 3, pp. 428–446.
Nuthall, Keith (2008). "Advantageous connections," *Modern Power Systems Magazine*, 5 September.
Nye, Joseph (1965). "Patterns and Catalysts in Regional Integration," *International Organization*, 19, 4.
Nye, Joseph (2003). *Understanding International Conflicts: An Introduction to Theory and History* (New York: Longman).
Nye, Joseph, ed. (1968). *International Regionalism: Readings* (Boston: Little, Brown).
Nye, Joseph S., and Robert O. Keohane (1971). "Transnational Relations and World Politics: An Introduction," in Keohane and Nye (eds.) *Transnational Relations and World Politics*, (Cambridge, MA: Harvard University Press), pp. xvi–xix.
Nygren, Bertil (2010). "Russia and Georgia—From Confrontation to War: What is Next?," in Roger E. Kanet, ed., *Russian Foreign Policy in the 21st Century* (Houndmills, Basingstoke, and New York: Palgrave Macmillan), pp. 107–108.
O Tuathail, Gearoid (2008). "Russia's Kosovo: A Critical Geopolitics of the August 2008 War over South Ossetia," *Eurasian Geography and Economics*, 49, 6, p. 670.
O'Neill, Michael (1996). *The Politics of European Integration: A Reader* (New York: Routledge).
Olcott, Martha Brill (2000). "Regional Cooperation in Central Asia and the South Caucasus," in Robert Ebel and Rjan Menon (eds.), *Energy and Conflict in Central Asia and the Caucasus* (New York: Rowman & Littlefield), pp. 129–134.
Olcott, Martha Brill (2006). "International Gas Trade in Central Asia: Turkmenistan, Iran, Russia, and Afghanistan," in David G. Victor, Amy M. Jaffe, and Mark H. Hayes (eds.), *Natural Gas and Geopolitics: From 1970 to 2040* (Cambridge: Cambridge University Press), p. 210.
Ozen, Cinar (1998). "Neo-functionalism and the Change in the Dynamics of Turkey-EU Relations," *Perceptions*, 3, 3.
Palosaari, Teemu (2004). *Beyond Normal Politics: Integration, EU Enlargement and Securitisation Theory* (Helsinki: Centre for European Studies).
Panfilova, Victoria (2008). "'Viza TRASEKA' pozvolit snizit' transportnye sbory i poshliny" ("Visa TRACECA" Will Facilitate the Reduction of Transportation Duties and Tariffs), *Ekho*, 17 March.
Papava, Vladimir (2005). "Georgia's Economic Role in the South Caucasus," *Problems of Economic Transition*, 48, 4, p. 89.
Paynter, John, M. E. Hawkesworth, and Maurice Kogan (1992). *Encyclopedia of Government and Politics* (New York: Routledge).
Peck, Anne E. (2004). *Economic Development in Kazakhstan: The Role of Large Enterprises and Foreign Investment* (New York: Routledge), pp. 193–194.
Petersen, Alexandros (2007a). "We must not keep the peace under the puppeteers of Putin," *European Voice*, 22 March.
Petersen, Alexandros (2007b). "Closer to Russia but more integrated with the West," *Turkish Daily News*, 6 September.

Petersen, Alexandros (2007c). "Integrating Azerbaijan, Georgia, and Turkey with the West: The Case of the East-West Transport Corridor," *CSIS Commentary*, 10 September.

Petersen, Alexandros (2008a). "The 1992-93 Georgia-Abkhazia War: A Forgotten Conflict," *Caucasian Review of International Affairs*, 2, 4, pp. 187–199.

Petersen, Alexandros (2008b). "Consolidating EU Energy Security: The East-West Transport Corridor and Western Integration," *Romanian Journal of International and Regional Studies*, 4, 2–3, pp. 20–23.

Petersen, Alexandros (2008c). "Regions in Between: Europe, NATO and the Geopolitics of Shifting Frontiers," *Turkish Policy Quarterly*, 7, 2, pp. 59–69.

Petersen, Alexandros (2008d). "Georgia: Next Step for Euro-Atlanticism?," *RUSI Newsbrief*, 18 February.

Petersen, Alexandros (2009a). "Russia May Score Final Coup in Energy Battle," *The Moscow Times*, 15 July.

Petersen, Alexandros (2009b). "Turkey's Multivector Energy Hub: Ignore at Your Own Peril," *Radio Free Europe/Radio Liberty*, 31 August 2009.

Petersen, Alexandros (2010). *Toward a U.S.-Azerbaijan Strategic Partnership* (Washington DC: Atlantic Council).

Petersen, Alexandros, and Taleh Ziyadov (2007). "Azerbaijan and Georgia: Playing Russian Roulette with Moscow," *Central Asia–Caucasus Analyst*, 10 January.

Petersen, Christian E., and Nina Budina (2003). "Governance Framework of Oil Funds: The Case of Azerbaijan and Kazakhstan" (Washington DC: World Bank Group).

Piebalgs, Andris (2007). "EU and Turkey: Together for a European Energy Policy." Speech, Conference on "Turkey and the EU," Istanbul, Turkey, 5 June.

Piebalgs, Andris (2008). "We Consider Trans-Caspian Pipeline Design an Important Issue," *Caspian Energy Journal*, 42, 3.

Pirani, Simon (2009). "Turkmenistan: An Exporter in Transition," in Simon Pirani (ed.), *Russia and CIS Gas Markets and Their Impact on Europe* (Oxford: Oxford University Press), pp. 271–296.

Pop, Valentina, (2008). "Energy Security: South Stream v. Nabucco," *SETimes.com*, 14 February, http://www.setimes.com/cocoon/setimes/xhtml/en_GB/features/setimes/features/2008/02/14/feature-02 last accessed 1 April 2009.

Poviliunas, Audrius (2006). "South Caucasus in the Context of European Neighbourhood Policy," *Lithuanian Foreign Policy Review*, p. 127.

Powell, Robert (1991). "Absolute and Relative Gains in International Relations Theory," *American Political Science Review*, 85, 4, pp. 1303–1320.

Powell, Robert (1993). "Absolute and Relative Gains in International Relations Theory," in David Baldwin (ed.), *Neorealism and Neoliberalism: The Contemporary Debate* (New York: Columbia University Press), p. 211.

Puchala, Donald (1981). "Integration Theory and the Study of International Relations," in Richard L. Merritt and Bruce M. Russett (eds.), *From National Development to Global Community: Essays in Honor of Karl W. Deutsch*(London: Allen and Unwin), p. 151.

Rehn, Olli (2007). *Keynote Speech to the Bosphorus Conference*, Istanbul, 5 October.

Rehn, Olli (2008). *Keynote Speech to the Bosphorus Conference*, Istanbul, 10 October.

Rengger, N. J. (2000). *International Relations, Political Theory, and the Problem of Order: Beyond International Relations Theory?* (New York: Routledge).

Richardson, J. J. (2002). *European Union: Power and Policy-making* (New York: Routledge).

Richardson, Jeremy John (2001). *European Union: Power and Policy-making* (New York: Routledge).

Rochester, J. Martin (2002). *Between Two Epochs: What's Ahead for America, the World, and Global Politics in the Twenty-First Century?* (Upper Saddle River: Prentice Hall).

Rosamond, Ben (2000). *Theories of European Integration* (New York: Palgrave).

Rosamond, Ben (2005). "The Uniting of Europe and the Foundation of EU Studies," *Journal of European Public Policy*, 12, 2.

Rosamond, Ben (2006). "The Uniting of Europe and the Foundation of EU Studies: Revisiting the Neofunctionalism of Ernst B. Haas," in *The Disparity of European Integration: Revisiting Neofunctionalism in Honour of Ernst B. Haas* (New York: Routledge), pp. 28–30.

Rotberg, Robert I. (2004). "Strengthening Governance: Ranking Countries Would Help," *Washington Quarterly*, 28, 1, pp. 71–81.

Rusbult, Caryl E., and Paul Van Lange (1996). "Interdependence Processes," in *Social Psychology: Handbook of Basic Principles* (New York: Guilford Press).

Russert, Bruce and John Oneal (2001). *Triangulating Peace: Democracy, Interdependence, and International Organizations* (New York: W.W. Norton).

Sadri, Mahmud (2001). "Iran: Paper analyses Russo-Iranian shared strategic interests, differences," *Hamshahri*, 11 March, http://w3.nexis.com/new/results/docview/docview.do?docLinkInd=true&risb=21_T15042308927&format=GNBFI&sort=BOOLEAN&startDocNo=51&resultsUrlKey=29_T15042308931&cisb=22_T15042308930&treeMax=true&treeWidth=0&csi=10962&docNo=53.

Schimmelfennig, Frank (1999). "The Double Puzzle of EU Enlargement," *ARENA Working Papers*, 15.

Schimmelfennig, Frank (2001). "The Community Trap: Liberal Norms, Rhetorical Action, and the Eastern Enlargement of the European Union," *International Organization*, 55, 1, pp. 47–80.

Schimmelfennig, Frank, and Ulrich Sedelmeier (2002). "Theorizing EU Enlargement: Research Focus, Hypotheses, and the State of Research," *Journal of European Public Policy*, 9, 4, pp. 500–528.

Schimmelfennig, Frank, and Ulrich Sedelmeier (2004). "Governance by Conditionality: EU Rule Transfer to the Candidate Countries of Central and Eastern Europe," *Journal of European Public Policy*, 11, 4, pp. 661–679.

Schleifer, Yigal (2007). "Questions Cloud Turkish-EU Energy Cooperation," *EurasiaNet*, 12 December.

Schleifer, Yigal (2008). "Turkey: Caucasus Crisis Leaves Ankara Torn Between U.S. and Russia," *EurasiaNet*, 9 November.

Schmitter, Philippe, and Ernst B. Haas (1964). *Mexico and Latin American Economic Integration* (Berkeley: University of California Institute of International Studies).

Schmitter, Philippe C. (2003). "Neo-Neo-Functionalism," in Antje Wiener and Thomas Diez (eds.), *European Integration Theory* (Oxford: Oxford University Press).

Schmitter, Philippe C. (1969). "Three Neo-Functional Hypotheses about International Integration," *International Integration* 23, 1, pp. 161–166.

Schneider, Volker, and Thierry Vedel (1999). "From High to Low Politics in Franco-German Relations: The Case of Telecommunications," in Douglas Webber (ed.), *The Franco-German Relationship in the European Union* (New York: Routledge).

Schreurs, Miranda Alice (2007). *The Environmental Dimension of Asian Security: Conflict and Cooperation Over Energy, Resources, and Pollution* (Washington DC: U.S. Institute of Peace).

Sedelmeier, Ulrich (2000a). "The EU and Eastern Enlargement: Risk, Rationality, and Role-Compliance," in Maria Green Cowles and Michael Smith (eds.), *The State of the European Union: Risks, Reform, Resistance and Survival*, 5, pp. 164–185.

Sedelmeier, Ulrich (2000b). "Sectoral Dynamics of EU Enlargement: Advocacy, Access, and Alliances in a Composite Policy," *Journal of European Public Policy*, 9, 4, pp. 627–649.

Shaffer, Brenda (2002). *Borders and Brethren: Iran and the Challenge of Azerbaijani Identity* (MIT Press: Cambridge), pp. 209–212.

Shaffer, Brenda (2004). "Iran's Internal Azerbaijani Challenge: Implications for Policy in the Caucasus," in Moshe Gammer (ed.), *The Caspian Region Volume 1: A Re-Emerging Region* (New York: Routledge), pp. 125–127.

Shaked, Haim, and Itamar Rabinovich (1980). *The Middle East and the United States: Perceptions and Policies* (Transaction Publishers).

Shuker, Stephen A. (2000). "The European Union: From Jean Monnet to the Euro," in Dean J. Kotlowski (ed.), *The European Union: From Jean Monnet to the Euro* (Athens: Ohio University Press, 2000), pp. 24–25.

Sidorenko, Alexy, and Arzu Geybullayeva (2010). "The Internet in Azerbaijan," and Alexy Sidorenko, "Internet, Society, and Democracy in Georgia," *Caucasus Analytical Digest* 15, pp. 4, 7.

Sinclair-Webb, Emma (2008), "Giving Turkey the Silent Treatment," *European Voice*, 11 December.

Siroky, David S., and David Aprasidze (2011). "Guns, Roses, and Democratization: Huntington's Secret Admirer in the Caucasus," *Democratization*, 18, 6, pp. 9–10, 12.

Skripkina, Olga (2008). Interview with General Secretary of TRACECA's Permanent Secretariat of the Inter-Governmental Commission Rustan Jenalinov, "Koridor, rozhdenyi dlya konkurentsii" [Corridor, born to be competitive], *RZD Partner*, 7 November 2008, http://www.rzd-partner.ru/interview/2008/11/07/333180.html.

Slider, Darrell (1997). "Democratization in Georgia," in Karen Dawisha and Bruce Parrott (eds.), *Conflict, Cleavage, and Change in Central Asia and the Caucasus* (New York: Cambridge University Press), pp. 185, 190, 193–194.

Smith, Gare (2004). "The BTC Pipeline Case Study: Following Through on Global Compact Commitments," United Nations Global Compact, 13 January 2004, http://www.unglobalcompact.org/docs/issues_doc/human_rights/Resources/Case_Story_BTC_Pipeline.pdf.

Smith, Gare (2006). "Human Rights and Security Monitoring Assessment of AGT Pipeline Projects in Azerbaijan," June 2006, http://www.bp.com/liveassets/bp_internet/bp_caspian/bp_caspian_en/STAGING/local_assets/downloads_pdfs/f/FH_Az_Vol_Princ_Assess_Final_en.pdf.

Smith, Gare (2007). "Human Rights and Security External Monitoring Assessment of the AGT Pipeline Projects in Azerbaijan," May 2007, http://www.bp.com/liveassets/bp_internet/bp_caspian/bp_caspian_en/STAGING/local_assets/downloads_pdfs/f/FH_VP_Assessment_Azerbaijan_2006_Final.pdf.

Smith, Gare (2008). "Human Rights and Security External Monitoring Assessment of the AGT Pipeline Projects in Azerbaijan," Foley Hoag LLP, 2008.

Smith, Karen E. (2005). "The Outsiders: The European Neighbourhood Policy," *International Affairs* 81, 4.

Socor, Vladimir (2003). "Too Little, but Not Yet Too Late," *Wall Street Journal*, 6 October.

Socor, Vladimir (2005). "Trans-Caspian Oil Pipeline Planned in Kazakhstan," *Eurasia Daily Monitor* (Washington DC: Jamestown Foundation).

Socor, Vladimir (2006a). "Azerbaijan Keeps Solidarity with Georgia Despite Russian Energy Supply Cuts," *Eurasia Daily Monitor* (Washington DC: Jamestown Foundation).

Socor, Vladimir (2006b). "Kyiv Changing Ideas, Mixing Signals on Odessa-Brody Oil Pipeline," *Eurasia Daily Monitor* (Washington DC: Jamestown Foundation).

Socor, Vladimir (2006c). "Odessa-Brody-Poland Oil Transport Project Shelved Indefinitely," *Eurasia Daily Monitor* (Washington DC: Jamestown Foundation).

Socor, Vladimir (2006d). "Azerbaijan Spearheading Initiative on Trans-Caspian Gas Pipeline," *Eurasia Daily Monitor* (Washington DC: Jamestown Foundation).

Socor, Vladimir (2007a). "Iran-Armenia Gas Pipeline: Far More Than Meets the Eye," *Eurasia Daily Monitor* (Washington DC: Jamestown Foundation).

Socor, Vladimir (2007b). "Azerbaijan at the Forefront of Caspian Basin Energy Initiatives," *Eurasia Daily Monitor* (Washington DC: Jamestown Foundation).

Socor, Vladimir (2007c). "Kars-Tbilisi-Baku Railroad: Azerbaijan as Locomotive of Regional Projects," *Eurasia Daily Monitor* (Washington DC: Jamestown Foundation).

Socor, Vladimir (2009). "No Gas Sources Foreseen for Gazprom's South Stream," *Eurasia Daily Monitor* (Washington DC: Jamestown Foundation).

Socor, Vladimir (2012). "Beijing Proposes Turkmenistan-China Gas Pipeline Through Northern Afghanistan," *Eurasia Daily Monitor*, 9, 116, 19 June.

Spanier, John (1990). *Games Nations Play* (Washington DC: CQ Press).

Spatharou, Angeliki (2001). "Geopolitics of Caspian Oil: The Role of Integration of the Caspian Region into World Economy in Maintaining Stability in the Caucasus," in Bulent Gokay (ed.), *The Politics of Caspian Oil* (New York: Palgrave), pp. 25, 28–29, 31–34.

Spero, Joan (1985). *The Politics of International Economic Relations* (New York: St. Martin's Press).

Starr, Harvey (2007). "Democracy and Integration: Why Democracies Don't Fight Each Other," *Journal of Peace Research*, 34, 2.

Starr, S. Frederick (2005). "The Baku-Tbilisi-Ceyhan Pipeline: School of Modernity," in Svante E. Cornell and S. Frederick Starr (eds.), *The Baku-Tbilisi-Ceyhan Pipeline: Oil Window to the West* (CACI: Washington DC).

Starr, S. Frederick (2010). Public remarks at "Does the U.S. have a Caspian Policy?" event, featuring C. Boyden Gray and Alexandros Petersen at Johns Hopkins University SAIS, Washington DC.

Stefano, Nikos, General Secretary of the Ministry of Development, (2007). 10th Athens Forum of the Energy Community, Athens, 24 April.

Stefanou, Constantin, and Helen Xanthaki (1999). "The Principle of the Effective Protection of the Individual in EC Law and the Dialectic of European Integration Theory," *Northern Ireland Legal Quarterly*, 50, 2.

Stephenson, Paul (2010). "Let's Get Physical: The European Commission and Cultivated Spillover in Completing the Single Market's Transport Infrastructure," *Journal of European Public Policy*, 17, 7, pp. 1039–1057.

Sterling-Folker, Jennifer (2000). "Competing Paradigms or Birds of a Feather? Constructivism and Neoliberal Institutionalism Compared," *International Studies Quarterly*, 44, 1.

Stern, Jonathan (2006). "Natural Gas Security Problems in Europe," *Asia Pacific Review*, 13, 1.

Stern, Jonathan, and Michael Bradshaw (2008). "Russian and Central Asian Gas Supply for Asia," in Michael Bradshaw, et al. (eds.), *Natural Gas in Asia: The Challenges of Growth in China, India, Japan, and Korea* (Oxford: Oxford University Press), pp. 260–262.

Stromberg, Roland (1992). *Europe in the Twentieth Century* (Englewood Cliffs: Prentice Hall).

Stulberg, Adam (2007). *Well-Oiled Diplomacy: Strategic Manipulation and Russia's Energy Statecraft in Eurasia* (Albany: SUNY Press).

Sulaberidze, Mariam (2008). "Global Information Society, Digital Divide, and Digital Opportunities of the South Caucasus Region," in *First Annual Students Scientific Conference: Development Perspectives of South Caucasus Region* (Tbilisi: Caucasus University, 2008).

Svota, Iveta, Mark S. Johnson, and Stephen Heyneman (2007). "Education and the Crisis of Social Cohesion in Azerbaijan and Central Asia," *Comparative Education Review* 51, 2, pp. 159–180.

Swanstrom, Niklas (2005). "China and Central Asia: A New Great Game or Traditional Vassal Relations?," *Contemporary China*, 14, 45, pp. 569–584.

Sweet, Alec Stone, and Wayne Sandholtz (1997). "European Integration and Supranational Governance," *Journal of European Public Policy*, 4, 3, pp. 297–317.

Swietochowski, Tadeusz (1995). *Russia and Azerbaijan: A Borderland in Transition* (New York: Columbia University Press), pp. 218–219.

Tassinari, Fabrizio (2006). "A Synergy for Black Sea Regional Cooperation," *CEPS Policy Brief*, 105 (Brussels: CEPS).

Taylor, Paul (2007). "EU, Turkey Play up Energy to Keep Accession Lights On," *Kathemirini*, 7 June.

Teimuraz Gorshkov, and George Bagaturia (2001). "TRACECA—Restoration of Silk Route," *Japan Railway & Transport Review*, 28, September.

Tekin, Ali, and Iva Walterova (2007). "Turkey's Geopolitical Role: The Energy Angle," *Middle East Policy*, 14, 1, pp. 84–94.
Telo, Mario, ed. (2001). *European Union and New Regionalism* (Aldershot: Ashgate).
Torbakov, Igor (2008). *The Georgia Crisis and Russia-Turkey Relations* (Washington DC: Jamestown Foundation).
Touboul, Bernard (2003). "The Central Asia Harmonisation of Border Crossing Procedures in the Perspective of TRACECA," Regional Seminar on Trade Facilitation and Customs Modernization, 4–8 August.
Tow, William (1990). *Subregional Security Cooperation in the Third World* (London: Lynne Rienner).
Toynbee, Arnold J. (1987) *A Study of History: Volume II* (Oxford: Oxford University Press).
Tromp, Hylke (1989). "Interdependence, Security, and Peace Research," in James Rosenau and Hylke Tromp (eds.), *Interdependence and Conflict in World Politics* (Brookfield: Avebury).
Tsagareishvili, Paata, and Gogita Gvenetadze (2009). "New Caspian Oil Production Will Bypass Russian Transport," *Oil and Gas Journal*, 104, 4, pp. 54–61.
Tsereteli, Mamuka (2009). *The Impact of the Russia-Georgia War on the South Caucasus Transportation Corridor* (Washington DC: Jamestown Foundation).
Tutuncu, Mehmet (1998). "Turkish Foreign Policy in the Caucasus," in Ole Hoiris and Sefa Martin Yurukel (eds.), *Contrasts and Solutions in the Caucasus* (Oakville CT: Aarhus University Press), p. 480.
van Apeldoorn, Bastiaan, Henk Overbeek, and Magnus Ryner (2003). "Theories of European Integration: A Critique," in *A Ruined Fortress? Neoliberal Hegemony and Transformation in Europe* (New York: Rowman & Littlefield), p. 21.
van der Leeuw, Charles (2000). *Oil and Gas in the Caucasus and Caspian: A History* (New York: St. Martin's Press), pp. 128–130, 134, 139.
Vanhercke, Bart (2006). "Political Spillover, Changing Advocacy Coalition, Path Dependency, or Domestic Politics?," Conference Paper, Governing Work and Welfare in an Enlarged Europe, May 19–20, 2006, http://www.ose.be/files/publication/bvanhercke/Vanhercke_2006_PaperConfWisconsinMay2005_Vers030506.pdf.
Verdun, Amy (2002). "Merging Neofunctionalism and Incrementalism," in Amy Verdun (ed.), *The Euro: European Integration Theory and Economic and Monetary Union* (New York: Rowman & Littlefield, 2002), pp. 11–13.
Vitale, Alessandro (2007). " The EU Wants to Build an Energy Strategy in the Caspian Region," Caucaz EuropeNews. 1 September, http://www.caucaz.com/home_eng/breve_contenu.php?id=293 , accessed 19 March 2009.
VoluntaryPrinciples.org (2009). "The Voluntary Principles Expands Membership, Renews Focus on Local Implementation," 24 March 2009, http://www.voluntaryprinciples.org/files/vp_oslo_plenery_press.pdf.
Wallace, Helen (2000). "EU Enlargement: A Neglected Subject," in Maria Green Cowles and Michael Smith (eds.), *The State of the European Union: Risks, Reform, Resistance and Survival*, 5, pp. 149–164.
Waltz, Kenneth (2008). *Realism and International Politics* (New York: Routledge).
Waltz, Kenneth N. (1986). "Anarchic Orders and Balances of Power," in Robert O. Keohane (ed.), *Neorealism and Its Critics* (New York: Columbia University Press), pp. 98, 100.
Waters, Christopher P. M. (2003). "Who Should Regulate the Baku-Tbilisi-Ceyhan Pipeline?" *Georgetown International Law Review*, 16.
Welz, Christian (2008). *The European Social Dialogue Under Articles 138 and 139 of the EC Treaty* (Leiden: Kluwer Law).
Wendt, Alexander (1999). *Social Theory of International Politics* (New York: Cambridge University Press).
Winrow, Gareth (1999). *Turkey and Caspian Energy* (Abu Dhabi: Emirates Center for Strategic Studies and Research), p. 9.

Winrow, Gareth (2000). *Turkey and the Caucasus: Domestic Interests and Security Concerns* (London: Royal Institute of International Affairs), p. 1, 3–5, 8–10, 15–16, 27–28.

Winrow, Gareth (2007). "Geopolitics and Energy Security in the Wider Black Sea Region," *Southeast European and Black Sea Studies*, 7, 2, pp. 217–235.

World Bank (2003). *Azerbaijan Building Competitiveness: An Integrated Non-Oil Trade and Investment Strategy (INOTIS), Volume 1: Summary Report* (Washington DC: World Bank Group).

World Bank (2005). *Azerbaijan: Issues and Options Associated with Energy Sector Reform* (Washington DC: World Bank Group).

World Bank and United Nations (2008). *Georgia: Summary of Joint Needs Assessment Findings Prepared for the Donors' Conference of October 22, 2008 in Brussels* (Washington DC: World Bank Group and United Nations).

World Bank/United Nations Report (2008). *Georgia: Summary of Joint Needs Assessment Findings*, December, http://siteresources.worldbank.org/INTGEORGIA/Resources/301645-1224598099977/GEJNA2008.pdf.

Yenikeyeff, Shamil (2009). "Kazakhstan's Gas Sector," in Simon Pirani (ed.), *Russia and CIS Gas Markets and Their Impact on Europe* (Oxford: Oxford University Press), pp. 345–347.

Yenikeyeff, Shamil Midhatovich (2008). "Kazakhstan's Gas: Export Markets and Export Routes," Oxford Institute for Energy Studies. November.

Zhukov, Stanislav (2009). "Uzbekistan: A Domestically Oriented Gas Producer," in Simon Pirani (ed.), *Russia and CIS Gas Markets and Their Impact on Europe* (Oxford: Oxford University Press), p. 389.

Zinin, Yuri N., and Alexei V. Maleshenko (1994). "Azerbaijan," in Mohiaddin Mesbahi (ed.), *Central Asia and the Caucasus After the Soviet Union: Domestic and International Dynamics* (Gainesville: University Press of Florida), pp. 109–111, 114.

Ziyadov, Taleh (2007). "Azerbaijan State Oil Company to Build Refineries Abroad," *Eurasia Daily Monitor* (Washington DC: Jamestown Foundation).

Ziyadov, Taleh (2007). "Azerbaijan" In S. Frederick Starr (ed.), *The New Silk Roads: Transport and Trade in Greater Central Asia* (Washington DC: Central Asia-Caucasus Institute).

Index

About the Author

Alexandros (Alexi or Alex) Petersen (1984–2014) was an American academic, writer, and geopolitical energy specialist. He wrote *The World Island: Eurasian Geopolitics and the Fate of the West* (Praeger: 2011), published over 120 articles, reports, and chapters in books, and made more than fifty media appearances internationally. Petersen received a BA in War Studies with First Class Honors from King's College London and an MSc and PhD in International Relations from the London School of Economics. He died January 17, 2014 in a Kabul bombing and shooting attack that killed twenty-one people.